STONEHENGE
DECODED

Other books by the author

Splendor in the Sky, Harper and Row, 1961.

The Sun and Its Planets, Holt, Rinehart and Winston, Young Owl Series, 1964. (With Fred Moore.)

Meteors, Comets and Meteorites, McGraw-Hill, Undergraduate Series in Astronomy, 1964.

The Life of a Star, Holt, Rinehart and Winston, Wise Owl Series, 1965. (With Fred Moore.)

Earth and Space Science, D. C. Heath, 1965. (With C. W. Wolfe, H. Skornik, L. J. Battan and R. H. Fleming.)

STONEHENGE
DECODED

GERALD S. HAWKINS
in collaboration with JOHN B. WHITE

A DELTA BOOK

This book is dedicated to Lord Snow of the City of Leicester, in pursuit of the two cultures.

A DELTA BOOK

Published by
DELL PUBLISHING CO., INC.
1 Dag Hammarskjold Plaza
New York, N.Y. 10017
Copyright © 1965 by Gerald S. Hawkins & John B. White
Delta ® TM 755118, Dell Publishing Co., Inc.

Reprinted by arrangement with Doubleday and Company, Inc.,
Garden City, N.Y.
Fifteenth Printing
Manufactured in the United States of America

ILLUSTRATION CREDITS

Plates 17, 18, 21, 22, 23. Photos from CBS Television Network
    program "Mystery of Stonehenge."
Plates 6, 7, 8, 9, 10, 11, 14, 15, 16. Photos by Stephen Perrin.
Plates 1, 2, 3, 4, and map on page v. Crown Copyright,
    Ancient Monuments Branch,
    Ministry of Public Building and Works, England.
Plates 5, 12, 13, 19. Photos by H. W. Edgerton.
Plate 20. Photo by C. A. Newham.

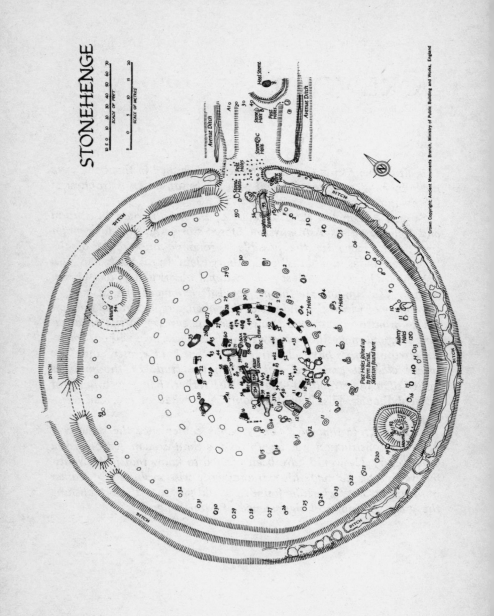

STONEHENGE

SCALE OF FEET
SCALE OF METRES

Avenue Ditch

Heel Stone

Avenue Ditch

DITCH

DITCH

DITCH

DITCH

DITCH

Mound 94

Aubrey Holes

Post Holes, joined up
to form burial.
Skeleton found here

Slaughter Stone

Altar Stone

"Z" Holes

"Y" Holes

FOREWORD

It is altogether fitting that the discoveries described in this book were made by an astronomer affiliated with the Smithsonian Astrophysical Observatory.

Samuel P. Langley, third secretary of the Smithsonian Institution and founder of its Astrophysical Observatory, was the first major scientist to recognize the possible astronomic importance of the "rude, enormous monoliths" of Salisbury Plain. In his book The New Astronomy he wrote, "Most great national observatories, like Greenwich or Washington, are the perfected development of that kind of astronomy of which the builders of Stonehenge represent the infancy. Those primitive men could know where the sun would rise on a certain day, and make their observation of its place . . . without knowing anything of its physical nature." By "that kind of astronomy" he meant classical positional observation, the study of the motions rather than the structures—the "where" rather than the "what"—of heavenly bodies. His "new astronomy" was what we now call astrophysics.

Langley wrote that in 1889, by happy coincidence the same year in which construction was begun on the Smithsonian Astrophysical Observatory. He would have been pleased to know that just seventy-five years after he made his extraordinarily wise evaluation a worker in the observatory which he founded would play a part in establishing the great astronomical significance of Stonehenge.

AUTHOR'S PREFACE

Every visitor to Stonehenge wonders in some way or other what its purpose could have been. The rugged stones are blank with no words of dedication, no constructional notation, and no readable clues. Because of this the word "decoded" needs some explanation.

As this book will show, there is a wealth of information in the positioning of the stones, in the successive master plans of the structure and in the choice of the site itself. There is much to read at Stonehenge without invoking ancient or modern words. It presents a unique cryptic puzzle, the solution of which has led to an understanding of the minds of prehistoric people. Before, with only vague legends to guide us, the remote past seemed incomprehensible. Now, perhaps, the door of prehistory stands ajar.

My working hypothesis has gradually developed over the past two years: If I can see any alignment, general relationship or use for the various parts of Stonehenge then these facts were also known to the builders. Such a hypothesis has carried me along over many incredible steps. In retrospect it is a conservative hypothesis for it allows the Stonehenger to be equal to, but not better than, me. Many facts, for example the 56-year eclipse cycle, were not known to me and other astronomers, but were discovered (or rather rediscovered) from the decoding of Stonehenge.

There can be no doubt that Stonehenge was an observatory; the impartial mathematics of probability and the celestial sphere are on my side. In form the monument is an ingenious computing machine, but was it ever put to use? As a scientist I cannot say. But in my defense a similar skepticism can be turned toward other probers of ancient humanity. Do we need to see lip marks on a drinking cup, blood on a dagger and sparks from a flint striking pyrites to convince us that these things were indeed used?

This investigation was carried out at the Smithsonian Astrophysical

Observatory, Harvard College Observatory, Boston University, and at the site of Stonehenge and the surrounding English countryside. It has led me into fields of the humanities as well as fields of science and in some measure I have crossed the bridge between the "Two Cultures" of Sir Charles P. Snow.

The work has brought me in touch with many people who have offered helpful advice and encouragement. Notably I would like to gratefully acknowledge discussions with R. S. Newall, H. Hencken, R. J. Atkinson, S. Piggott, H. E. Edgerton, A. Thom and C. A. Newham. My wife Dorothy has maintained a keen interest in Stonehenge and the meaning of the various discoveries. I am grateful to Mr. F. Friendly and the staff of CBS for placing on permanent record the astronomical events at Stonehenge that took place in 1964 and which might otherwise, like the events of previous millennia, have passed according to schedule but unseen.

The book would not have been possible without the untiring assistance and encouragement of John B. White. Mrs. Edith Homer typed the various draft copies and the final manuscript efficiently and uncomplainingly.

GERALD S. HAWKINS

Maugus Hill
Wellesley Hills, Massachusetts
February, 1965

This printing contains much new archaeological information kindly provided by Professor Atkinson.

G.S.H

May, 1966

COLLABORATOR'S NOTE

Being neither astronomer nor archaeologist, I was able to contribute to this book only an intense, amateur interest in Stonehenge, and some research into its history—real and imaginary.

JOHN B. WHITE

Cambridge, Massachusetts
February, 1965

CONTENTS

LIST OF ILLUSTRATIONS

LIST OF PLATES

Chapter 1

THE LEGENDS

Stonehenge is unique. In all the world there is nothing quite like the gaunt ruin which Henry James said "stands as lonely in history as it does on the great plain." Immense and still, it seems beyond man, beyond mortality. In its presence, within those silent circles, one feels the great past all around. One can almost see and hear . . . until one tries to imagine precisely *what* sights and sounds animated that place, what manner of men moved there, in that inconceivably remote past when it was new.

What was it? What purpose did it serve, this monument and memorial of men whose other memorials have all but vanished from the earth? Was it a city of the dead? A druid place of horrid sacrifice? A temple of the sun? A market? A pagan cathedral, a holy sanctuary in the midst of blessed ground? *What was it . . . and when?*

There have been many stories and legends about the strange place, and some of those legends cling to it still.

Stonehenge was so old that its true history was probably forgotten by classic times. Greek and Roman writers hardly mention it. When the practical Roman invaders came to Britain they paid it little reverence—after all, Rome had her temples, and Egypt her pyramids, in better condition, perhaps, than this group of stone blocks. Indeed, there is evidence that the Romans may have knocked chips off of some of the blocks—they may have considered the place a possible center for revolutionary activity.

Not until the Dark Ages brought back mystery did the old stones begin to stir men's fancies. By then any clear memory of the origin and use of the "gigantick pile" had long since evaporated. It was necessary to create for it a biography, almost as one in those credulous days patched together lives for the innumerable blessed and unrecorded saints.

We cannot know who the first such biographer of Stonehenge was. It may have been the sixth-century Gildas, whom some have called "the Wise" and some have said never even existed. It may have been Aneurin the great Welsh bard, who in the seventh century allegedly sang of the beginnings of that work of giants. It may have been the ninth-century Nennius, who wrote romantically of a stone memorial erected for British nobles treacherously slain—but was that memorial Stonehenge, and was there really a monk-chronicler named Nennius?

We do know that by the twelfth century it was well wrapped in speculation and legend. Wace, the Anglo-Norman, said it was called "hanging stones" in both English and French—"Stanhengues ont nom en Englois, pierres pendues en Francois"—and Henry of Huntingdon explained that the name was well deserved, because the stones "hang as it were in the air." (Others have thought the epithet referred not to the stones but to the criminals who may have hung from them.) Henry did not think "Stanhengues" was Britain's greatest marvel, however. The first wonder of the land, he wrote, was a "wind which issues from a cavern in the earth at a mountain called Pec" (medievalists may know where Mount Pec is—I don't); the second marvel was Stonehenge, "where stones of a wonderful size have been erected after the manner of doorways, so that doorway seems to have been raised upon doorway, nor can any one conceive by what art such great stones have been raised aloft, or why were there constructed." Giraldus Cambrensis, friend of Richard Coeur de Lion and of John I, also classified the stones as a marvel, as did most of the other chroniclers of that time.

The attempt to account for the origins of that marvel resulted in myths. Those myths were most effectively gathered together and passed on by that master historian and myth-dispenser of the twelfth century, Geoffrey of Monmouth.

I shall quote from Geoffrey at some length, not because I am a legend-lover—I'm not—but because this one old myth, so well related by him, continued to be the source for most of the fabling about Stonehenge for five hundred years.

According to Geoffrey (*Histories of the Kings of Britain*),* the story of Stonehenge began in the time of "King Constantine," when "a certain Pict that was his vassal . . . feigning that he did desire to

---

* In order to keep the undergrowth of footnotes pruned to a minimum, references to works cited throughout this book have been tucked into the general bibliography at the end.

hold secret converse with him, when all had gone apart, slew him with a knife in a spring-wood thicket." Then Vortigern, Earl of the "Gewissi," was "panting to snatch the crown," but Constantine's son Constans was made king, so Vortigern "hatcheth treason": he bribed the Picts and "made them drunken" so that they "burst into the sleeping-chamber, and fell suddenly upon Constans . . . smiting off his head."

Vortigern then became king.

Soon there was trouble. ". . . three Brigantines . . . arrived on the coasts of Kent full of armed warriors and captained by the two brethren Horsus and Hengist. . . ."

(Actually, Hengist and Horsa did lead the first Saxon invasion of England, in the fifth century. Apparently Vortigern "covenanted" with the Saxons and married Hengist's daughter Rowen, but Hengist continued to pursue a course of "subtle craft." According to Bede and the *Anglo-Saxon Chronicle* the Saxons were given the island of Thanet but fought with their British hosts. Horsa was killed but Hengist and his son Aesc conquered the whole kingdom of Kent. As Geoffrey tells the story, it was done by deepest villainy.)

Having "made ordnance unto his comrades that every single one of them should have a long knife hidden along the sole of his boot," Hengist called a meeting of Britons and Saxons near Salisbury "on the Kalends of May," and "when . . . the hour had come . . . the Saxons set upon the princes that stood around" and "cut the throats of about four hundred and sixty. . . ."

(The legends become badly confused here. Some declare that it was Vortigern who betrayed the British "princes." In any case, there was much strife between Britons and Saxons. It was in a battle between them at "Mount Badon" (Bath? Badbury?) in the sixth century that King Arthur was first mentioned; Nennius listed him in passing as a "dux bellorum," or leader of warriors, of the Britons; not for many decades thereafter did he become an outstanding semi-mythical hero. A British king, Ambrosius Aurelianus, who may have existed—if so, he was probably of Roman descent—was supposed in a few of the legends to have been Arthur's magical father Uther Pendragon. The modern town of Amesbury is thought by some to have derived its name from Ambrosius. Geoffrey wrote that Ambrosius was Uther Pendragon's brother, and ruled with the help of the wizard Merlin.)

One day the king came to Salisbury, "where the earls and princes

lay buried whom the accursed Hengist had betrayed," and was
"moved to pity and tears began to flow . . . at last he fell to pon-
dering . . . in what wise he might best make the place memorable
. . . the green turf that covered so many noble warriors."
Merlin said,

"If thou be fain to grace the burial-place of these men with a
work that shall endure forever, send for the Dance of the Giants
that is in Killaraus [Kildare?], a mountain in Ireland. For a structure of stones is there that none of this age could raise save his
wit were strong enough to carry his art. For the stones be big, nor
is there stone anywhere of more virtue, and, so they be set up
round this plot in a circle, even as they be now there set up, here
shall they stand for ever."

The king burst out laughing and said, "But how may this be, that
stones of such bigness and in a country so far away may be brought
hither, as if Britain were lacking in stones enow for the job?" Merlin
answered, "Laugh not so lightly . . . in these stones is a mystery, and
a healing virtue against many ailments. Giants of old did carry them
from the furthest ends of Africa and did set them up in Ireland
what time they did inhabit there . . . not a stone is there that lacketh
in virtue of witchcraft. . . ."

The king was convinced. "The Britons . . . made choice of Uther
Pendragon, the king's brother, with fifteen thousand men, to attend
to this business." The armada put to sea "with a prosperous gale."
The Irish heard of the proposed seizure of their monument, and King
Gilloman raised a "huge army," vowing that the Britons should not
"carry off from us the very smallest stone of the Dance." But the invaders "fell upon them straightway at the double-quick . . . prevailed
. . . pressed forward to mount Killaraus. . . ."

Then the would-be monument-movers were faced with the problem of how to transport those great stones. "They tried huge hawsers
. . . ropes . . . scaling ladders [memories of the lists of weapons in
Caesar's *Gallic Wars!*] . . . never a whit the forwarder. . . ." Merlin
had to take over. He "burst out laughing and put together his own
engines . . . laid the stones down so lightly as none would believe
. . . bade carry them to the ships," and they all "returned unto Britain with joy" and there "set them up about the compass of the burial-
ground in such wise as they had stood upon mount Killaraus . . . and
proved yet once again how skill surpasseth strength."

Geoffrey added that Uther Pendragon, and King, or Emperor, Constantine, were both buried at Stonehenge.

Most of Geoffrey's story is useful only as entertainment, but there are certain bits of it that merit consideration, or if not consideration at least comment. ITEM: Stonehenge was certainly not built to commemorate either Saxon or British dead—but it is interesting that the old legend so firmly links it with such a use, when it was only recently found to have been a place of burial. ITEM: Geoffrey said that its stones were of supreme "virtue." It is true that there was general reverence for the mystic powers of stones for a long time after the coming of Christ—in 452 A.D. the Synod of Arles denounced those "who venerate trees and wells and stones" and such denouncement was repeated by Charlemagne and others down to recent times—but modern discoveries, to be discussed later, have demonstrated the possibility that the stones of Stonehenge may have been regarded by their original erectors as of especially sovereign powers. Two stones were crucial in the legend of Arthur: the unknown lad became king by literally one twist of the wrist—he grasped that mysterious sword and "lightly and fiercely pulled it out of the stone"—and then the only man, or being, who could have saved him became "assotted and doated on one of the ladies of the lake . . . that hight Nimue . . . and always Merlin lay about the lady to have her maidenhood, and she was ever passing weary of him, and fain would have been delivered of him, for she was afeared of him because he was a devil's son . . . and so on a time it happed that Merlin showed to her in a rock whereas was a great wonder . . . so by her subtle working she made Merlin to go under that stone to let her wit of the marvels there, but she wrought so there for him that he came never out for all the craft he could do," and—Merlin thus entombed beneath that stone—the fate of king and kingdom was sealed. ITEM: Geoffrey's statement that the stones had come to Ireland from Africa is understandable when we remember that Africa was regarded as the home of strangeness; man-of-affairs–writer Pliny declared in the first century A.D., "Out of Africa always something new." The legend that the stones had been set up in Ireland may not be so absurd as it might seem. It is quite possible that stones as big and sacred as those of Stonehenge might have been set up in ritual arrangement and then moved from place to place. (The present theory as to where they probably did come from will be discussed in Chapter 4.) And certainly "fifteen thousand men" could have been used in any such moving. ITEM: It is interesting that in the legend Merlin did not resort to simple magic to whisk the stones from the old site to the new. He was of course more than capable of that; legendizers other than Geoffrey state that he transported the stones by his "word of power" only. Could it be that there lurks folk-memory of actual moving of those stones in that story of Merlin's "engines"?

In the realm of purer myth, there may be more than engineering connection between Merlin and Stonehenge. Some mythographers have thought that the name "Merlin" is a corruption of the name of the ancient Celtic sky god "Myrddin," who might have been worshiped at stone monuments. A Welsh triad states that the whole of Britain, before men came, was called "Clas Myrddin," or "Merlin's Enclosure." The Welsh folklorist John Rhys in an 1886 Hibbert Lecture said, "I have come to the conclusion that we cannot do better than follow the story of Geoffrey, which makes Stonehenge the work of Merlin Emrys, commanded by another Emrys, which I interpret to mean that the temple belonged to the Celtic Zeus, whose later legendary self we have in Merlin." In 1889 Professor A. T. Evans wrote in the *Archaeological Review* that Stonehenge was an advanced representation of sepulchral architecture, "where the cult or worship of departed ancestors may have become associated with the worship of the Celtic Zeus; the form under which the divinity was worshipped would have been that of his sacred oak."

Whatever the truth, if any, hidden in the legend of Merlin's building Stonehenge, that legend dominated the field for centuries. For some reason—because the stones were actually there, and therefore defied complete mythologizing?—Merlin's Salisbury Plain effort did not feature heavily in the fables about King Arthur and his Table Round. But among the stories which the late Middle Ages fed on concerning the marvelous life and times of the real monument, that account which credited Stonehenge to Merlin was the most popular. And as Arthur faded into the land of faery, the story of "how Merlin, by his skill and magic's wondrous might,/from Ireland hither brought the Sonendge in a night" (Michael Drayton, in the poem *Polyolbion*), began to arouse practical curiosity. Geoffrey's tale and its many variants fell into disrepute.

The anonymous fifteenth-century author of the *Chronicle of England* stated briskly that he didn't believe that Merlin had put up those stones. In the next century Polydore Vergil, archdeacon of Wells, not crediting Merlin, wrote that the monument, "made of great square stones, in form of a Crown," had been raised by "the Britains" in memory of Ambrosius. The Elizabethan historian-anti-quary William Camden had no heart for speculating about the origin of the "huge and monstrous piece of work," remarking sadly that

"Our countrymen reckon this for one of our wonders and miracles; and much they marvel from whence such huge stones were brought . . . for my own part . . . I am not curiously to argue and dispute, but rather to lament with much grief, that the authors of so notable a monument are thus buried in oblivion. Yet some men think them to be no natural stones hewn out of the rock, but artificially made of pure sand, and by some gluey and unctuous matter knit and incorporated together . . . and what marvel? Read we not, I pray you, in Pliny, that the sand or dust of Puteoli, being covered over with water becometh a very stone?"

Spenser of course found Geoffrey's exotic tale much to his taste. In *The Faerie Queene's* "chronicle of Britons kings,/from Brute to Uthers rayne./And rolles of Elfin Emperours,/till time of Gloriane," he told how Constantine "oft in battell vanquished/Those spoilefull Picts, and swarming Easterlings" but was "annoyd with sundry bordragings/Of neighbour Scots, and forrein Scatterlings" before "Vortigere/Usurpt the crowne" and "sent to Germanie, straunge aid to reare . . ." Hengist and Horsa, "well approu'd in warre . . . making vantage of their civill jarre . . . grew great . . ." and Vortigern was "enforst the kingdome to aband." With the help of his son Vortimer the king was restored to power, whereupon "*Hengist* seeming sad, for what was donne,/Received is to grace and new accord,/Through his faire daughters face, and flatring word;/Soone after which three hundred Lordes he slew/Of British bloud, all sitting at his bord;/Whose dolefull moniments who list to rew,/Th'eternall markes of treason may at *Stonheng* vew."

Less poetic theorizers tended to agree that the "dolefull moniment" had been erected in post-Roman times, but not by Merlin.

In the seventeenth century, men suddenly became interested in everything. The new scientific spirit, which John Donne apprehensively noted "throws all in doubt," left nothing unconsidered. Those geniuses, near-geniuses and ordinary men of an extraordinary time focused their attention on all things both great and small. Newton was something of an alchemist. Wren, the geometrician-astronomer and architect, was also a pioneer in the practice of blood transfusion.

Hooke invented or claimed to have invented almost as many imaginative devices as Leonardo da Vinci.†

Naturally, something as strange as Stonehenge did not escape such curious-minded persons' attention. Many people visited the site, and many more wrote about it.

Early in the century the king, James I, visited Stonehenge. He was so excited by it that he ordered the celebrated architect Inigo Jones to draw a plan of the stones and find out how the structure had come into being. Jones apparently inspected Stonehenge, but unfortunately for us he left no direct record. All we know is that in 1655 his son-in-law John Webb published a book, *The Most Remarkable Antiquity of Great Britain, vulgarly called Stone-Heng, Restored*, in which he gave the gist of what he described as "some few undigested notes" left by Jones. The book is a stirring demonstration of what happens when a master craftsman attacks a problem in his field without having access to the facts. Inigo Jones looked at Stonehenge with an architect's eyes, considered it as an architectural puzzle, and produced some architecturally oriented conclusions that were as closely reasoned as they were—inevitably—wrong. His book is a fascinating document, a perfect gold mine of perceptive observation, shrewd analysis, miscellaneous information (not all of it erroneous) and first-rate lore-based logic. (Fig. 1.)

Jones praised the monument for the "rarity of its invention . . . beautifull Proportions," pronounced it "elegant in Order . . . stately in aspect," and proceeded to examine the credentials of various of the candidates who had been named as possible builders of the edifice. One-two-three he ticked them off:

"Concerning the Druid's . . . certainly, Stoneheng could not be builded by them, in regard, I find no mention, they were at any time either studious in architecture, (which in this subject is chiefly to be respected) or skilful in any thing else conducing thereunto. For, Academies of Design were unknown to them: publique Lectures in the Mathematiques not read amongst them: nothing of their Painting, not one word of the Sculpture is to be found, or scarce any Science

† As an example of the range of interest of those first children of science, here are a few of the listings of A Century of the Names and Scantlings of such Inventions, As at present I can call to Mind . . . , a book published by the Marquis of Worcester in 1663: "a ship-destroying engine . . . unsinkable ship . . . sea-sailing fort . . . pleasant floting garden . . . to and fro lever . . . portable bridge . . . needle alphabet . . . most conceited tinderbox . . . artificial bird . . . pocket ladder . . . flying man . . . imprisoning chair . . . semi-omnipotent engine . . . stupendious water work." The Marquis spent so much money trying to develop some of his "scantlings" that he finally went broke.

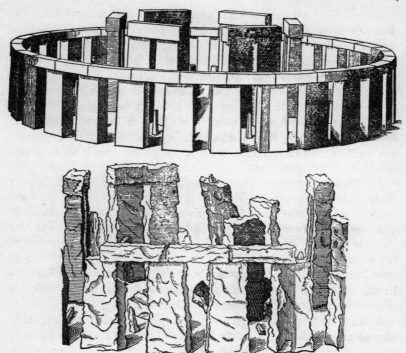

Fig. 1. A rendering of Stonehenge, from *Stone-Heng, Restored*, presumably by Inigo Jones, showing the monument as he imagined it to be.

(Philosophy and Astronomy excepted) proper to inform the judgement of an Architect . . ."

As for the early "Britans," they were "savage and barbarous people, knowing no use at all of garments . . . destitute of the knowledge . . . to erect stately structures, or such remarkable works as Stoneheng. . . .

"In a word therefore let it suffice, Stoneheng was no work of the Druid's, or of the ancient Britans; the learning of the Druid's consisting more in contemplation then practice, and the ancient Britans accounting it their chiefest glory to be wholly ignorant in whatever Arts. . . ."

Finally, "as for that ridiculous Fable, of Merlins transporting the stones out of Ireland, it's an idle conceit."

Having thus disposed of those candidates for the honor of having erected the "work built with much Art, order and proportion," Jones produced his own candidates. "Considering what magnificence the Romans in prosperous times anciently used in all works . . . their knowledge and experience in all Arts and Science: their powerfull means for effecting great works: together with their Order in building, and the manner of workmanship accustomed among them, Stoneheng in my judgement was a work, built by the Romans, and they the sole Founders thereof. . . . But if it is objected, If Stoneheng a Roman work, how comes it, no Roman Author makes mention of it? I answer, their Historians used not to commit to writing any particular work or action the Romans performed: if so, how vast would their volumes have been?"

For their architectural style the builders "in all likelihood . . . for so notable a structure as Stoneheng, made choice of the Tuscane rather than any other order, not only as best agreeing with the rude, plain, simple nature of those they intended to instruct . . . but also . . . to magnifie to those then living the virtue of the Auncestors for so noble an invention."

When was it built? "Happily, about the times, when the Romans having setled the Country here . . . reduced the naturall inhabitants of this Island unto the Society of Civill life. . . ."

And its use? It was "originally a Temple . . . sacrifices anciently offered at Stoneheng . . . were Buls or Oxen, and severall Sorts of beasts, as appears by the heads of divers kinds of them, not many years since there digged up." As a temple it was dedicated to the sky god, Coelus—because it stood in an open plain, under the sky, because it was circular, like the round earth, and because its stones were shaped like flames and fire was the celestial element.

The diligent and admirable Inigo ended his sturdy attempt to date Stonehenge architecturally with this most engaging benediction:

"Whether, in this adventure, I have wafted my Barque into the wished Port of Truths discovery concerning Stoneheng, I leave to the judgement of Skilfull Pilots. I have endevoured, at least, to give life to the attempt, trending perhaps to such a degree, as either may invite others to undertake the Voyage anew, or prosecute the same in more ample manner, in which, I wish them their desired Successe, and that with prosperous Gales they may make a more full and certain discovery."

Often, since I have set out on the same voyage of discovery concerning Stonehenge, I have felt the warmth of that 300-year-old wish, and added to it my own good wishes for the "desired Successe" of future investigators of the old mystery.

There had been other seventeenth-century theories about Stonehenge. A certain Edmund Bolton had in 1624 credited it to the famous Boadicea, or Boudicca, a British queen who led a great revolt against Rome but was defeated and took poison. Her name in Celtic means something like "Victoria." Wrote Bolton,

"The story of Bunduca [Boadicea] . . . was so little understood by Monmouth, as it doth not appear at all . . . higher than to Her no Books do reach . . . and the profound oblivion which covers the Author, and the first intention of rearing them [the Stonehenge stones], where now they still defie the weather, doth strongly fortifie my suspition, that the stones were consecrated to the Glory of Bunduca, and of her Captains slain in her quarrel, so long time since as Nero Caesars dayes. . . ."

But the Jones theory, as advanced by his son-in-law Webb, stirred the most controversy.

In 1663 Dr. Walter Charleton, one of the notable physicians who attended Charles II, disputed Webb in a tract with the resounding title of *Chorea Gigantum, or the most famous Antiquity of Great-Britain, vulgarly called* STONE-HENG, *Standing on Salisbury Plain, Restored to the* DANES. A resounding effort indeed. In the full flood of that same ample prose which had but lately been applied to the King James version of the Bible, Dr. Charleton began,

"Your Majesties Curiosity to survey the subject of this discourse, the so much admired Antiquity of STONE-HENG . . . sometime . . . so great and urgent, as to find a room in Your Royal Breast, amidst Your Weightiest Cares . . . animated Me, to make strict Enquiry into the Origin and Occasion of the Wonder (so the Vulgar call it) so far as the gloomy darkness of Oblivion would admit . . . [of] that Gigantick Pile, whose dead Remains . . . sleeping in deep Forgetfulness, and well-nigh disanimated by the Lethargy of Time (which often brings the River Lethe to flow as well aboveground, as below). . . ."

He then gave his opinion:

"Having diligently compared STONE-HENG with other antiquities of the same kind . . . in Denmark . . . I now . . . conceive it to have been Erected by the Danes, when they had this Nation in subjection; and principally, if not wholly Design'd to be a Court Royal, or place for the Election and Inauguration of their Kings; according to a certain Strange Custom, yet of eldest Date. . . ."

Dr. Charleton's diligence was praised by poet "Rob. Howard"—
"How much obliging is your learned Care!/Still busie to pursue, or to
repair . . ."—and his theory was applauded by none other than John
Dryden:

> . . . you may well give
> to Men new vigour, who make Stones to live.
> Through you, the DANES (their short Dominion Lost)
> A longer conquest that the Saxons boast.
> Stone-HENG, once thought a Temple, you have found
> A Throne, where Kings, our Earthly Gods, were crown'd. . . .

But Charleton's claim, that the Danes were "the Authors of this
Stupendious Building, that doth so amaze and amuse its beholders,"
was given short shrift. Webb immediately reiterated father-in-law
Inigo's Roman-origin theory, and others entered the polite but
spirited controversy.

An odd effusion called A Fool's Bolt Shot at Stonehenge, ascribed
to one John Gibbons who flourished in the 1670s, asserted that it was
"an old British triumphal tropical temple, erected to Anaraith, their
goddess of victory, in a bloody field there won by illustrious Stanenges
and his Cerngick giants, from King Divitiacus and his Belgae."

The two great diary-keepers, John Evelyn and Samuel Pepys, both
visited the site, and reported typically. Evelyn, interested in natural
history and in architecture, wrote (July 22, 1654), "After dinner . . .
we passed over the goodly plain, or rather sea of carpet . . . arrived at
Stonehenge, indeed a stupendous monument, appearing at a distance like a castle. . . ." He thought that the "so many and huge
pillars of stone" had perhaps been parts of a "heathen . . . natural
temple," and he went on to state that "the stone is so exceedingly
hard, that with all my strength with a hammer could not break a
fragment, which hardness I impute to their so long exposure. . . ."
Pepys, more interested in people and affairs, wrote (June 11, 1668),
"Come thither, and found them as prodigious as any tales I ever
heard . . . God knows what their use was!"

In that time, however, there was what seems to have been the first
careful on-the-site investigation of the ancient monument in its history. John Aubrey is now remembered chiefly (if at all) for his collection of rambling biographies called Brief Lives, but a more solid
fame could be claimed for him: he was the first archaeologist, or
proto-archaeologist, of England. Camden and others had written of
antique sites, but they had drawn their information from records,
and usually confined their observations to secondhand description.

Even Inigo Jones approached Stonehenge more as an architect than an antiquary. Aubrey went to the site and poked around and measured.

He was born quite near Stonehenge, at a hamlet named Easton Pierse some thirty miles north of the monument, in 1625 or 1626. He grew up in what he termed "an Eremeticall solitude," which he disliked—"twas a great disadvantage to me in my childhood"—but which may have been a factor in the forming of a "strong and early impulse to Antiquitie . . . I was inclin'd by my Genius, from my Childhood to the Love of Antiquities and my Fate dropt me in a Country most suitable for such Enquiries." In particular, "Salisbury-Plaines, and Stonehenge I had known from eight years old. . . ."

Aubrey was anything but thorough. He started many large projects and finished none—*Brief Lives* (including the celebrated vignette of Shakespeare, "His father was a Butcher, and I have been told . . . that when he was a boy . . . when he kill'd a Calfe he would doe it in a high style, and make a Speech. . . .") existed only as a jumble of notes when he died. He confessed that he "wanted patience to go thorough Knotty Studies," and Anthony à Wood, the sour author of *Athenae Oxonienses*, called him "roving and magotie-headed." But Aubrey cut something of a figure in his time. He was a member of the Royal Society and a friend of the king and other important people, and his views were not without influence. And those views, in matters archaeological, were based on careful observation. With no evidence other than the stones themselves to reason from, he reasoned valiantly enough concerning the origin of Stonehenge. In 1663 he "tooke a Review" of the monument for Charles II, sketched it with commendable care (and his usual roving spirit—in one margin there appears, drawn with as much attention to detail as characterizes the outlines of the stones, a "batter-dasher"), and concluded:

"There have been several Books writt by learned men concerning Stoneheng, much differing from one another, some affirming one thing, some another. Now I have come in the Rear of all by comparative Arguments to give a clear evidence these monuments [he had also looked at other monuments, which will be discussed later] were Pagan Temples; which was not made-out before: and have also, with humble submission to better judgements, offered a probability, that they were Temples of the Druids. . . .

". . . my presumption is, That the Druids being the most eminent Priests, or Order of Priests, among the Britaines; 'tis odds, but that these ancient monuments . . . were Temples of the Priests of the most eminent Order, viz. Druids, and . . . are as ancient as those times. This Inquiry, I must confess, is a gropeing in the Dark; but although I have not brought it into a clear light; yet I can affirm that I have brought it from an utter darkness to a thin mist, and have gone further in this Essay than any one before me. . . ."

Aubrey was right about Stonehenge being more ancient than Roman or Saxon times, and possibly right about it having at some time served as a druid temple, but probably not right in his implied assumption that the druids built it. John Aubrey did much good work at Stonehenge, but his linking of the monument to the druids was a doubtful service.

There *were* druids. And they did come to Britain. But did they come before Stonehenge existed, or even when it was new? Were they its high priests? We do not know—but the evidence now is quite strong against that supposition.

There is, however, such strong and continuing interest in these glamorous, over-romanticized beings, and so much misunderstanding concerning their possible connection with Stonehenge, that a discussion of what is actually known about the druids seems in order here, to set the record straight.

The druids were the holy men, medicine men, teachers and judges of the Celts. Classic literature abounds in references to them. Caesar's account in *Gallic Wars*‡ is the most straightforward:

"Throughout Gaul there are two classes of persons of definite account and dignity. As for the common folk, they are treated almost as slaves. . . . One consists of druids, the other of knights. The former are concerned with divine worship, the due performance of sacrifices, public and private, and the interpretation of ritual questions: a great number of young men gather about them for the sake of instruction and hold them in great honour. . . . It is they who decide in almost all disputes . . . and if any crime has been committed or murder done, or if there is any dispute about succession or boundaries, they also decide . . . of all these druids one is chief . . . it is believed that their rule of life was discovered in Britain and transferred thence to Gaul. . . .

"Report says that in the schools of the druids they learn by heart a great number of verses, and therefore some persons remain

‡ Translated by H. J. Edwards, The Loeb Classical Library, 1917. Reprinted by permission of the Harvard University Press.

twenty years under training . . . they make use of Greek letters
. . . the cardinal doctrine which they seek to teach is that souls do
not die, but after death pass from one to another . . . besides this,
they have many discussions as touching the stars and their movement, the size of the universe and of the earth. . . . The whole
nation of the Gauls is greatly devoted to ritual observances, and
for that reason those who are smitten with the more grievous maladies and who are engaged in the perils of battle either sacrifice
human victims or vow so to do, employing the druids as ministers
for such sacrifices. They believe, in effect, that, unless for a man's
life a man's life is paid, the majesty of the immortal gods may
not be appeased . . . others use figures of an immense size, whose
limbs, woven out of twigs, they fill with living men and set on fire,
and the men perish in a sheet of flame. They believe that the
execution of those who have been caught in the act of theft or
robbery or some crime is more pleasing to the immortal gods; but
when the supply of such fails they resort to the execution even of
the innocent. . . .

"The Gauls affirm that they are all descended from a common
father, Dis, and say that this is the tradition of the druids. For
that reason they determine all periods of time by the number, not
of days, but of nights, and in their observance of birthdays and the
beginnings of months and years day follows night."

(Dis was the god of the dark underworld; the term "fortnight" still
bespeaks the custom of measuring time by nights rather than days.)

Pliny described the druids more romantically. He professed to de-
spise "Art Magicke," as he called it, but he respected its possible
powers, and he thought it his duty to set forth its history, and the
history of those who practiced it. The quotation is taken from the
1601 Holland translation of Pliny because the archaic language seems
best to fit the thought—that edition is the one from which Shakespeare probably drew material for some of the marvels Othello described to Desdemona:

"The sundrie kinds of magicke . . . execrable acts . . . may be
practiced after various sorts . . . for it worketh by means of Water,
Globes of Balls, Aire, Stars, Fire-lights, Basons and Axes. . . . The
follie and vanitie of Art Magicke . . . entermingled with medicina-
ble receits and religious ceremonies, the skill of Astrologie and arts
Mathematicall . . . in the realm of Persia, it found first footing,

and was invented and practiced there by Zoroastes . . . 5000 years before the War of Troy." [Actually, Zoroaster, or Zarathustra, lived in Persia about 600 B.C.]

Pliny said that Orpheus, Pythagoras, Empedocles and Plato "were so far in love" with the magic art that they "undertook many voyages" for its sake, and "this art they blazed abroad and highly praised." He said Moses also was a magician. Then the art came to "Fraunce," and there

"continued untill our daies: for no longer is it agoe than since the time of Tiberius Caesar, that their Druidae (the Priests and Wise Men of France) were by his authoritie put downe, togither with all the pack of such physicians, prophets, and wizards. But what should I discourse any longer in this wise, of that Art which hath passed over the wide ocean also, and gone as far as any land is to be seene, even to the utmost bounds of the earth; and beyond which, there is nothing to be discovered but a vast prospect of Aire and Water, and verely in Britaine at this day it is highly honoured, where the people are . . . wholly devoted to it. . . .

"The Druidae . . . esteeme nothing more sacred in the world, than Misselto, and the tree whereupon it breedeth, so it be on Oke . . . they may seeme well enough to be named thereupon Dryidae in Greeke, which signifieth . . . Oke-priests [the Greek word for "oak" was "drus" and Pliny's etymology may have been correct] . . . Misselto . . . they gather . . . very devoutly and with many ceremonies [when the] . . . moon be . . . just six daies old (for upon that day they begin their moneths and new yeares, yea and their severall ages, which have their revolutions every thirtie yeares) because shee is thought then to be of great power and force sufficient. . . . They call it in their language All-Heale, (for they have an opinion of it, that it cureth all maladies whatsoever) and when they are about to gather it, after they have well and duly prepared their sacrifices and festivall cheare under the said tree, they bring thither two young bullocks milke white . . . the priest arraied in a surplesse or white vesture, climbeth up into the tree, and with a golden hook or bill cutteth it off, and they beneath receive it . . . then fall they to kill the beasts . . . mumbling many oraisons & praying devoutly . . . now this persuasion they have of Misselto thus gathered, That what living creature soever (otherwise barraine) doe drinke of it, will presently become fruitfull thereupon . . . so vaine and superstitious are many nations in the world. . . ."

Pliny's conclusion is revealing, and damning for those present apologists who aver that "magicians" like the druids were harmless:

"See how this Art . . . is spread over the face of the whole earth! . . . the benefit is inestimable that the World hath received by the great providence of our Romans, who have abolished these monstrous and abominable Arts, which under the shew of religion murdred men for sacrifices to please the gods; and under the colour of Physicke, prescribed the flesh to be eaten as most wholesome meat."

Good loyal imperialist Pliny! Thus castigating foreign dietary abominations, he did not see fit to mention at this point the possibly embarrassing fact that in his city, in his time, "our Romans" were not innocent vegetarians; it is elsewhere in his voluminous writings that he chides Roman epileptics "who drinke the verie bloud of Fencers and Sword-plaiers as out of living cups" and deplores the cannibalism of "others that lay for the marow-bones, the very braine also of young infants, and never make straunge to find some good meat and medicine therein."

Dio Chrysostom, a contemporary of Pliny, had this to say of the druids: "It is they who command, and kings on thrones of gold, dwelling in splendid palaces, are but their ministers, and the servants of their thought."

Perhaps with time druids softened their customs and became more humane. Later accounts of them stress their wisdom, healing and teaching ability, and their judging. Their mystical powers were described as less savagely dependent on human sacrifice: they raised magic mists, cast "enervating spells," prophesied and in general attended to the ritualistic life of the people without demanding blood —or so say the accounts. It is always hard to find out about pagan priesthoods like the druids because so much of the literature about them has been filtered through Christian transmission.

The best present estimate is that the druids came with the Celts to Britain in about the fifth century B.C., and soon became the most influential priestly cult in the land. For centuries they were powerful. Indeed, they survived as priests, judges, doctors and educators, particularly of the royal young, after the Christians came to Britain in about the third century. More than six hundred years later Alfred the Great translated warnings against those who were "prone like beasts . . . baleful" in the following of "all this druidcraft."

Savage or benign, the druids were most picturesque. And the memory of them was never lost. In the seventeenth century interest in them revived. Samuel Butler in his satiric poem *Hudibras* scoffed at their belief in immortality—"Like money by the Druids borrow'd,/ In th'other world to be restor'd . . ."—but in general they inspired respectful curiosity. They still do. In 1781 a group calling itself "The Most Ancient Order of Druids" was established in London, and still flourishes. This group regards "Druidry" as more mystical and philosophical than religious, and lays claim to ancient, arcane wisdom inherited from semimythical people like the inhabitants of the lost continent, Atlantis.

These modern "Druids" have somehow established in the official mind so firm a conviction that they have legitimate connection with Stonehenge that they are allowed to conduct unauthentic ceremonies there on midsummer day at sunrise as if they really were re-enacting traditional rites. It is a pity, because this carrying-out of made-up "rituals" by a group which has no real knowledge of what the ancient druids thought or did—and no proof that they existed when Stonehenge was new—only confuses the ignorant and annoys the serious students of the past.

It is possible that the druids, the real druids, had something to do with Stonehenge when it was operative. Many things are possible. But it now seems extremely unlikely. One can but regret that John Aubrey gave the druids-built-Stonehenge theory such credence, because that theory has generated a distorted picture of Stonehenge as a ghastly place dedicated to human sacrifice and other frightful rituals presided over by white-robed priests with bloody hands. There *may* have been sacrifices at Stonehenge—we have no proof pro or con—but such sacrifices, if they took place, very probably were not directed by druids, since druids very probably were not present in England then, and such sacrifices were certainly not the only rituals practiced at the site.

Aubrey was a careful investigator and a fairly restrained theorizer. He would doubtless be amazed if he could return to see what his championing of the druids had grown into.

The seventeenth century was generally rather sober in its speculations about Stonehenge. Not so the next century. That supposedly restrained and neoclassic period produced some remarkably fanciful Stonehengerie. Opinions were advanced crediting it to most of the previously postulated originators with new candidates added including the Phoenicians.

In 1740 Dr. William Stukeley, renowned for his assistance in reconstituting the Society of Antiquaries which James I had abolished because of suspected politicking, published his notable, fascinating *Stonehenge, a temple restored to the British Druids.* Stukeley was a vigorous mixture of reckless imagination and meticulous investigation. He backed Aubrey's druid theory with all his might, so spiritedly that scholars think he should actually be given the greater share of the credit (?) for the subsequent popularization of that unfortunate notion, and he added an astonishing detail of his own; he declared that not only had the druids definitely worshiped at Stonehenge—what they had worshiped there had been the serpent! Stonehenge and similar stone circles, he claimed, had been serpent temples, or "Dracontia." He traced a lively "patriarchal history, particularly of Abraham," which continued with the "deduction of the Phoenician colony into the Island of Britain, about or soon after his time; whence the origin of the Druids . . . ," and credited his ancestral heroes with phenomenal powers: ". . . our predecessors, the Druids of Britain . . . advanced their inquiries, under all disadvantages, to such heights, as should make our moderns ashamed, to wink in the sun-shine of learning and religion."

When not occupied with his vaulting druidism, Stukeley did much good observational work at Stonehenge, however. He carefully measured distances between positions and tried to show that the builders had used a unit of length which he called a "druid cubit," a distance of 20.8 inches. He is credited with the first mention of the Avenue, which runs northeasterly from the monument, and he seems to have been the discoverer of the Cursus, a large low earthwork slightly to the north. And—surprisingly, for that still superstitious time—he tried to apply science to the dating of the monument. In what authorities think is the first recorded attempt to use laboratory methods to solve an archaeological problem he assumed that his druid builders had used the magnetic compass, and by comparing Stonehenge orientations with the rate of change of magnetic variation (a rate somewhat trickier to chase back over the centuries than he realized), he deduced that the date of building had been about 460 B.C. He was of course hopelessly wrong, but it was a brave try.

Stukeley was an energetic combination of subjective and objective reasoner. He succeeded in both confusing and clarifying the situation.

His *Stonehenge* . . . contains many passages of nostalgic charm. He later came almost to identify himself with his mystical priests in their "serpent temple," barely managing to keep one foot in the

eighteenth century. Stonehenge delighted him, in a non-druid fashion
—"it cannot but be the highest pleasure imaginable to a regular mind,
to walk round and contemplate the stately ruins. . . ." Indeed, he
seems to have feared that those ruins might not long outlast him:
". . . I have sketched the following prospects, taking in the county
almost round the circumference of the horizon. This use there will
be in them further; if it ever happen, that this noble work should be
destroyed: the spot of it may be found, by these views."

His work is especially interesting to astronomers, because it contains the first known reference to what has since become the most
famous single fact about Stonehenge, the fact that ". . . the principle
line of the whole work, [points to] the northeast, where abouts the
sun rises, when the days are longest." That fact is of crucial importance to understanding the nature of Stonehenge and will be discussed throughout the rest of this book.

In 1747 an architect of Bath, John Wood, outdid Stukeley. He published a book, *Choir Gaure, Vulgarly called Stonehenge, on Salisbury
Plain, Described, Restored, and Explained* . . . , which "explained"
things so succinctly that one might have thought no further explana-
tions would ever have been required. (Choir Gaure or Gawr has since
been supposed to mean "great" or "circular" "temple" or "gathering
place," although one interpreter, a Dr. John Smith who will be mentioned again shortly, believed that "choir" was the choir of a church
and "gaur" was derived from the same root as the word "caper" or
"he-goat.") Wrote Wood,

> "Caesar! even Julius Caesar, the high priest of Jupiter, and of
> Rome herself, undeniably proves the Brittanick Island to have been
> enriched with the great school of learning . . . wherein the Druids
> of the western world could perfect themselves in their profession
> . . . the venerable and stupendious work on Salisbury Plain, vul-
> garly ascribed to Merlin, the Prophet . . . appeared to me to be
> the remains of a Druidical temple . . . externally, of the real Mo-
> nopterick kind . . . neither could I avoid concluding, that the
> Britons and Hyperboreans were one and the same people. . . ."

(Greek and Roman poets and writers from Homer through Pliny
referred to a far northern people called "Hyperboreans." There will
be a discussion of these references in Chapter 8.) Wood then re-
capitulated classic accounts of a mostly mythical British king named
Bladud, whom he made synonymous with other legendary figures
named Aquila and Abaris. Bladud, he declared, ruled in Britain, then

"travelled into Greece for improvement at the very time when Zoroaster flourished in Persia, and Pythagoras . . . in Greece." There he
became famous for uttering oracles and building temples, including
the "Delphick temple itself." Finally Bladud-Aquila-Abaris returned
to Britain and founded the druid order. Stonehenge was erected by
priests of that order some time between then, about the fifth century B.C., and the birth of Christ. Wood also had a theory about
where the structure's stones had come from. He thought they had
been brought not from Ireland, by Merlin's skill, but from Marlborough Downs, just to the north of Stonehenge.

Soon after Wood wrote, a minister, William Cooke, Rector of Old-
bury and Didmarton in Gloucestershire, agreed with the architect's
theory and amplified it. "The vulgar opinion of its having been
raised by Aurelius Ambrosius . . . is scarce worth confuting," he
declared—Stonehenge had been erected by the druids, before Christ.
But the druids were so morally high-minded that they were not very
different, ethically, from Christians. Indeed, as Cooke stated—
drawing from sources which he did not reveal—"for the perpetual es-
tablishment and support of it [Stonehenge], they [the druids] were
wont to dedicate the tenth of all their substance." The rector approved of the Stonehenge columns, because Moses had built "an
altar and twelve pillars"; he approved of its circles, because a circle
is the "apposite emblem of that infinity which is applicable only to
the Supreme Being"; and he supposed that the druid stones, "these
Petrae Ambrosiae," were properly sanctified—"stones consecrated or
anointed with oil of roses."

In 1771 astronomy was invoked, apparently for the first time since
Stukeley, to account for the orientation of Stonehenge. Dr. John
Smith, identified chiefly as "the Inoculator of the Small-Pox," published a pamphlet titled *Choir Gaur, the Grand Orrery of the
Ancient Druids*. An orrery, named for the Earl of Orrery, was a
clockwork mechanism made to show planetary motions; Dr. Smith
maintained that Stonehenge was a numerical-mystical kind of calendar. For example, he supposed that since one of the monument's
circles had 30 stones, and since there are 12 "signs" of the ancient
zodiac, the 30 times 12 equalled 360, the number of days in the
"antient solar year." Amid his mysticizing Smith did repeat—and im-
prove on—Stukeley's concrete observation that the monument's principal axis was aligned to midsummer sunrise. As he phrased it,
Stonehenge was so laid out that when it was new, at dawn on midsummer day (the longest day of the year), the "Arch Druid standing

in his stall, and looking down the right line of the Temple . . . sees the sun rise. . . ."

Dr. Johnson, famous for his mighty pronouncements on most things in heaven and earth, did not neglect Stonehenge. Writing to Mrs. Thrale on October 9, 1783, he made this judicious observation: "It is in my opinion, to be referred to the earliest habitation of the island, as a druidical monument of, at least, two thousand years; probably the most ancient work of man upon the island. Salisbury Cathedral and its neighbour Stonehenge are two eminent monuments of art and rudeness, and may show the first essay and the last perfection in architecture."

In 1796 a Wiltshire clothier named Henry Wansey returned to the astronomical aspect of the monument with this report: "Stonehenge stands in the best situation possible for observing the heavenly bodies, as there is an horizon nearly three miles distant on all sides. But till we know the methods by which the ancient druids calculated eclipses with so much accuracy, as Caesar mentions, we cannot explain the theoretical use of Stonehenge." A most interesting point! There will be much more discussion of the possible use of Stonehenge as an eclipse predicter later in this book.

The nineteenth century, beginning with the Gothic romanticism of Byron, Shelley, Keats and the others, made much of the obligingly picturesque old ruin. Guidebooks, almost invariably describing Stonehenge as a druid temple, proliferated. Artists painted it, sinister beneath dark skies. People visited, and shivered in pleasant fright, and chipped away souvenir pieces. Whereas an earlier guidebook had deplored the "unaccountable Folly of Mankind in breaking pieces off with great Hammers," in the 1800s such hammers were rented by neighboring merchants for the specific purpose of chipping off mementoes, and for those too lazy to chip their own pieces, "unheeding shepherds of the plain will be ready to provide them . . . for . . . a few halfpence." It is fortunate that, as Evelyn noted, the stones of Stonehenge do not yield easily to would-be demolishers, otherwise such vandalism might have left little of it standing. Its popularity might literally have been its downfall.

That century also brought some of the most varied Stonehenge speculation.

In 1812 the antiquary Sir Richard Colt Hoare produced a fairly accurate chart of the Stonehenge positions as they actually were, not as he supposed they might have been. He dug vigorously in the surrounding area, although not in the Stonehenge enclosure itself, and proved that—as Stukeley had already noted—some of the ancient burials around the monument had been placed after Stonehenge was built; he found pieces of Stonehenge stones in certain burial pits. Sir Richard wrote,

"It is a melancholy consideration, that at a period when the sciences are progressively advancing, and when newly-discovered manuscripts are continually drawn forth from their cloistered retreats to throw a light on the ancient records of our country, it is mortifying, I say, that the history of so celebrated a monument as Stonehenge should still remain veiled in obscurity. The Monks may boldly assert, that Merlin and only Merlin was the founder of our temple; and we cannot contradict, though we may disbelieve. The revolution of ages frequently illustrates history, and brings many important facts to light; but here all is darkness and uncertainty; we may admire, we may conjecture; but we are *doomed* to remain in ignorance and obscurity."

But others did not agree that the obscurity enshrouding Stonehenge's secrets was doomed to remain total. In 1839 John Rickman, Fellow of the Royal Society, produced the opinion that the monument's stones must have been erected rather recently because, he stated, they gave evidence of having been worked with steel tools. In 1847 the Rev. H. M. Grover disagreed with that modern-origin theory; in *A Voice from Stonehenge* he supposed that the building had been done in the Saturnian or Golden Age, by "the might of a giant brood, which preceded in this, as in the Holy Land, the race of degenerate mortals of our own poor standard." He added that the work had apparently been directed by Egyptian architects, and druids.

Two years later the Hon. Algernon Herbert entered the lists with *Cyclops Christianus; Or, an Argument to Disprove the Supposed Antiquity of the Stonehenge and other Megalithic Erections in England and Brittany*. In his opinion, Stonehenge was not a sepulchre, although it was "erected in a vast and ancient cemetery," and it was built in the fifth century A.D.; because the necessary scientific abilities to create such a structure were lacking in Roman Britain but present as soon as the Romans left, the monument was "in considerable progress in 429, or 21 years after the independence of the island. . . ."

In the mid-1800s a diligent researcher named Henry Browne, of Amesbury, produced an ". . . Unprejudiced, Authentic, and Interesting Account which that Stupendous and Beautiful Edifice Stone-

henge, in Wiltshire, is Found to Give of Itself." He began by remarking that the old story of "Jeffry of Monmouth" was "almost too
absurd to merit even mentioning," went on to sum up various other
theories, noted in passing that since "Stonehenge stands not on the
summit, but on the gentle declinity of a hill" it could hardly have
been an astronomical observatory, and—reasoning from "considerations hitherto unnoticed"—reached the novel conclusion that the
Stonehenge stones had been erected during the days of Adam and
knocked down by the Flood. "Shall we . . . attribute their erection
to Britons, to barbarians?—silly thought!" As evidence, he adduced
the Biblically derived information that the lives of the antediluvians
were, "generally speaking, ten times the duration" of ours, they were
"both of greater stature and of greater strength," they "constructed
abodes . . . were conversant with . . . art; made instruments of music
. . . worked both in brass and in iron . . . erected places of worship
. . . ," and finally "they had continually before their eyes for more
than half the duration of the Antediluvian world, the presence of the
miraculously created Adam himself." He credited the planning of the
work to druids, and offered as clinching proof of his hypothesis the
fact that most of the Stonehenge dilapidation is now on the southwest side:

> ". . . to judge of the operation of the waters of the Deluge, we
> should conceive them, on issuing out of the bowels of the earth,
> to acquire such an elevation as, on the principle of gravity, would
> be sufficient to carry them over the countries which they were
> destined to inundate . . . the waters of the Deluge advanced against
> Stonehenge . . . from . . . the south-west. . . ."

In 1860 the *Quarterly Review* voiced the opinion that "it is little
wonder that sober-minded people look on the solution as hopeless,"
but people, sober-minded and otherwise, continued to speculate. In
1872 the eminent architect-scholar Sir James Fergusson reaffirmed the
old Saxon-origin theory. He had contemplated antiquities and "rude
stone monuments" all the way from Persepolis and Nineveh to the
British Isles and he thought Stonehenge was what Geoffrey of Monmouth had said it was, a sepulchre for the victims of Hengist's treach-
ery. In 1873 the Rev. L. Gidley ventured some astronomical observations, which have since been confirmed, and credited Dr. Smith
with having made the observation, previously noted, about the Stonehenge axis pointing to midsummer sunrise. And some time in that
decade an antiquary now identifiable only as "Dr. John Thurnam"

wrote a paper reviving the seventeenth-century theory of Edmund Bolton that Stonehenge, that "admirable monument," was the "place of Boadicea's buriall." According to his ingenious theory, the "dumbness of it speakes that it was not the work of the Romans, for they were wont to make stones vocall by inscriptions . . . that Stonage was a work of the Britaines the rudeness itself persuades. . . ."

In 1876 one W. Long theorized that Stonehenge was "inseparably connected" with the burials around it and supposed it to have been built by the Belgae, possibly with the help of Phoenicians. The next year Professor Nevil Story Maskelyne gave it as his opinion that the bluestones had come not from Ireland but from the Corstorphine Hills near Edinburgh; as for the sarsen stones [both of these types of Stonehenge stones will be discussed later], he did not know where they might have come from, but he felt that they "are capable of speaking in a language that has no ambiguity if we know how to interpret it. . . ."

Also in that decade there was carried out the first really accurate charting of Stonehenge. Jones, Aubrey, Wood, Smith, Colt Hoare and others, including Sir Henry James—not the author—and one Hawkshaw—not the detective—had mapped the site with precision varying from yards to feet or at best inches; in the 1870s W. M. Flinders Petrie, later an outstanding Egyptologist, produced a chart accurate to about an inch. Petrie thought that most of Stonehenge had been built before the Roman invasion, but that a few stones had been erected later, to the memory of Aurelius Ambrosius, Uther and Constantine, "and probably other chiefs, buried at intervals at Stonehenge." As for Merlin's alleged part in the project, Petrie wrote, "There is nothing of which a modern contractor need be ashamed. He [Merlin] is only said to have used 'the engines that were necessary' to remove the stones from Ireland to the ships, and they were brought over in the most matter-of-fact manner." But, he cautioned, "what is now necessary, to settle this much-disputed subject, is careful digging. . . ."

Neither Petrie's opinion nor his admonition ended the dispute. In 1883 one W. S. Blacket introduced a new element into the theorizing by announcing that everybody else was wrong—the creators of the mysterious structure had been not Britons, not Saxons, not Romans, not Merlin, not druids, not emigrants from Bible lands, not, in fact, men from any known lands. He concluded that those responsible had been the beautiful and marvelous (and mythical) people of the Lost Continent, Atlantis . . . via the New World. "The Apalacian Indians,

with their priests and medicine men, must have been the builders of
Stonehenge . . . [which] attests the truthfulness of Plato when he
brings into western Europe a great conquering people from beyond
the Pillars of Hercules." Plato did of course write about Atlantis in
the *Timaeus* and the *Critias*. He said that Solon had said that the
Egyptians had said that nine thousand years before there had been a
great island in the Atlantic, "larger than Libya and Asia together,"
which traded, flourished, grew proud and sent its "mighty host . . .
insolently advancing to attack the whole of Europe, and Asia to boot."
Then Fate intervened—"there occurred portentous earthquakes and
floods, and in one grievous day and night . . . the island of Atlantis
was swallowed up by the sea and vanished. . . ." Plato spent a lot of
time describing the political practices of the Atlanteans, and obviously meant the "island" to be understood as only a rhetorical device,
no more real than his "republic." Geologists agree that there has been
no vast earth convulsion as recently as 10,000 B.C., and even if there
had been, no land mass as large as the "island" Plato described could
have sunk far beneath the sea in one day and night. But the, or a,
Lost Continent still has believers. As undersea exploration shows
more and more of the Atlantic to be bare of evidence of past civilization therein inundated, the legend moves west; quite popular now
among the drowned lands fraternity seems to be Atlantis' Pacific
counterpart, the equally lost continent of Lemuria, or Mu.

A year after Blacket had invoked Atlantis, another cogitator, named
T. A. Wise, produced the last of the truly imaginative nineteenth-
century speculations about Stonehenge. He thought that it had been
one of the "high places of the Druids"—until it fell into the hands of
Buddhist missionaries.

One of the first reasonable Stonehenge theories after Petrie's was
stated by the son of an astronomer, John Lubbock, son of Sir John
William Lubbock. The father was noted for his work on comet or-
bits, eclipses and the moon-tide relation, and the son produced the
very good estimate that Stonehenge and similar stone relics were prod-
ucts of the Bronze Age, 1500–1000 B.C. His antiquarian work was so
highly regarded even in his own day that in 1900 he, like his father,
was elevated, and given the fitting title of Lord Avebury.

And so it went, the legendizing and theorizing and speculating,
until the end of the century. Many people wondered about Stonehenge, but nobody really knew anything about the origins and history of the mysterious place. Everybody was free to speculate; many
did; hardly any theory was untried. Along with the Atlanteans and

other ethereal folk, various more plebeian real races such as the Celts and Phoenicians and Belgae were again put forward as the builders.

In the midst—or mist—of the theorizing there was growing a conviction that there should be more archaeologically directed investigation of the site. And early in the present century such investigation began. The digging and the identifying and the dating has gone on with increasing enthusiasm ever since, and now much of the mystery of Stonehenge has been cleared away.

The poets have thought this something of a shame. Yeats especially clung to the old mysteries, and druidism. He steadfastly maintained that there was more than a touch of druid in him, and in "these fitful Danaan rhymes" he sang of "a Druid land, a Druid time" and the blessed druid paradise of Tir-na-nOg. He never gave up his "little bag of dreams."

But as far as Stonehenge is concerned, the replacing of dreams by reality may prove beneficial to the dreamers as well as to the scientists; the new archaeological discoveries, to be discussed later, are revealing so much that is astonishing, and, in a new way, picturesque.

The "when" of Stonehenge is now known to be long before Saxons, Danes, and even before the Romanized Britons. The "how" of the massive structure—how those great stones were assembled and erected—has not been so definitely established, but is providing much thought for archaeologists, engineers and others interested in the capabilities of early men. The "why" of Stonehenge is one of the main subjects of this book.

Chapter 2

THE PEOPLE

Who *did* build Stonehenge?

Amateur delvers into the past of the British Isles usually form the impression that it was to Ireland that the earliest, and certainly the most romantically named, Dawn People came. There is so much literature about Ireland's first families—and how remote, improbable, and picturesque they sound: Partholonians, Fomorians, Nemedians, Fir Bolg, Tuatha de Danann, Milesians, Dravidians.

Actually, England was explored as soon as Ireland, or sooner: it is 250 miles closer to Continental Europe. Any exotic semimythical people who really did make their way to Erin probably passed through Albion en route. Unfortunately, the Romans conquered England and broke the thread of bardic narrative of what had been before. What cared the hard-bitten legions for local gossip?

The Britons were quickly Romanized and within a generation were talking and thinking Latin. A workman scratched SATIS ("enough") on a tile at the end of the day. (Just such a tile, dated back to 50 A.D., was found recently.) But Ireland was never imperialized. Her bards and monks handed down uninterrupted the old stories. Odd old stories, to be sure, obviously more poetry than truth. Personally I have little use for legends. I much prefer the hard facts. But some authorities think there is enough indicated fact in the fancy to make the myths worthy of our attention. For it is sure that prehistoric colonizers of Ireland also touched England, and Scotland, as well as the other "holy Island to the west." The traits and talents of the prehistoric men of Ireland must have been the same as those of other peoples in the British Isles. What were those traits and talents? Let the *Book of Conquests* and the other old manuscripts speak.

There are mentioned three waves of early invaders; the Fomorians, the Sons of Partholan, and the Nemedians, not necessarily in that order, and not all necessarily pre-Stonehenge.

The Fomorians, fierce and dangerous, were "gloomy sea-giants . . .

warlike . . . very troublesome to all the world." They were also diligent farmers; "they made sheep land." And—keep in mind the construction of Stonehenge—"they built towers." They brought their
skills from Africa, by way of what we now call Spain. The Partholonians also seem to have come from Spain. Not so much is recorded
of their habits, except that they fought with the fierce Fomorians,
more successfully than they did with Fate. "Plague buried them . . .
and the land was waste thirty years." The Nemedians came from
Greece, via Scythia, and brought political skill. When the "gloomy
sea-giants" oppressed them they sent back to the "nobles of Greece"
for help. Their plea must have been most persuasive. Help soon came
in the form of "an immense host of warriors, with Druids and Druid-
esses" and also—one cannot but be curious as to what kernel of fact
there might be in *this* flight of fancy—"venomous animals . . . hurtful strange animals. . . ." What animals could be hurtful and
strange, brought all the way from Greece? Dogs?

Thus reinforced, the Nemedians "overcame the towers of the Fomorians," and prospered, until a "great wave" came from the sea
and "drowned and annihilated" conquerors and vanquished alike.
(One is tempted to think that the "great wave" was caused by the
flooding of the North Sea across the land bridge from England to the
mainland, but that event took place much earlier, perhaps as early
as 10,000 B.C., when the last glaciers melted, and it was a very slow
process of gradual flooding.) Some Nemedians survived the deluge,
but "downcast and fearful of the plague" they departed for England
and for their old home, Greece. "And the land was desert for the
space of two hundred years."

Then came the Fir Bolg, people who seem to have had characters
as unexotic as their name was strange. According to the legends, they
originated in Greece, as peasants working for those ancestors of Homer's "well-greaved" Achaeans who centuries later burnt the topless
towers of Ilium. The Fir Bolg were industrious and competent farmers, with the praiseworthy custom of creating fertile fields out of
wasteland by the laborious process of covering it with soil carried in
bags. "They made clovery plains of the rough-headed hills with the
clay from elsewhere." The legend-makers thought that their leather
bags gave them their name, "Fir Bolg" being interpreted to mean
"men of the bags." As did the Egyptians with the hard-working Israelites, however, their masters grew demanding beyond endurance.
And the Fir Bolg, "tired, weary and despondent," threw off their
"intolerable bondage . . . made canoes and fair vessels of the skins

and rope bags for carrying the earth," and sailed away. They reached
Ireland in one week, according to the Irish sagas. One week, for a
journey in skin canoes, 1600 miles from Greece to the Pillars of
Hercules and 1100 miles farther to Ireland; a total of 2700 miles in
seven days . . . those were "fair vessels" indeed.

Once they had arrived safely in their new home, the men of bags
immediately set to work carrying soil again, to make more fertile the
green hills. They seem never to have stopped moving dirt, which is
an interesting trait when one comes to consider the digging processes
necessary for construction of monuments like Stonehenge. Actually,
people on the Isle of Aran were doing just this in the present century. Flaherty's prize-winning documentary film shows the men claw-
ing soil from cracks in the rock on that wind-swept spot off the coast
of Ireland. The soil was carried to garden patches for potato growing.

Next came the most endearing and attention-worthy of all those
legendary folk, the mystical Tuatha de Danann. Their name seems to
have meant People or Children of Dana, Dana being their god, although some mythologists link them to the moon goddess Danaë.
The Tuatha were memorable for charm equaled only by wide wisdom.
At first they had lived "in the northern isles of Greece," and they
were very learned. "They knew lore and magic and druidism and
wizardry and cunning . . . and surpassed the sages of the arts of
heathendom in lore and science . . . diabolic arts . . . every sort of
gentilism. . . ." Perhaps "gentilism" included diplomacy, because for
a while the Tuatha "went between the Athenians and the Philis-
tines," apparently as mediators. They seem to have been good ar-
rangers of other people's affairs, as well as of their own.

According to the legend, the Tuatha had descended from those
Nemedians who had returned to Greece. The Tuatha sailed away "in
speckled ships" to reclaim their heritage. "They came with a great
fleet to take the land from the Fir Bolg." They landed on that ritual
day, the first of May, traditional day for the combat of winter and
summer, and overcame the carriers of soil in bags.

The *gentil* wizards ruled for a time, in a sort of Golden Age of
benevolence, with "lore and science" that most assuredly would have
been of capital value to the great work on Salisbury Plain.

Then came the most numerous, best-organized of the legendary
fortune hunters.

Milesius was "standing on his rooftop one day in a far distant land . . . contemplating and looking over the four quarters of the world," and lo! he saw "a shadow and likeness of a land and lofty island far away." Naturally, he "brought his ships on the sea," to that land, and after a "bad welcome" routed the Tuatha de Danann, to whom in return for temporal power he granted immortality. The wizards went "beneath the happy hills . . . to live forever." (Ireland still reveres her People of the Hills.)

The Milesians brought their share of legends. One story credits them with causing the notorious serpentless condition of Ireland, thus: a Milesian forebear was cured of snake bite by none other than Moses (snake-handling ran in Moses' family—Aaron's rod changed into a serpent before Pharaoh). Then Moses promised the Milesian that his people would come to a "fertile land never to be defiled by snakes." One of the Milesians married Pharaoh's daughter Scota, who gave her name to Ireland,* and later their descendants went out of Egypt to Spain and thence, all those legendary centuries before anti-ophidian Patrick, to Ireland. Many stories of the Milesians are no more than bedtime stories, but as usual in the old accounts there are to be found in the mists of legend those little definite details which indicate plausibility: the Milesians had "federations of aristocratic republics" and political unification. They carried out a "consistent foreign policy." In the bardic arts they were supreme; their bards could remember twelve books, along with 350 kinds of poetic meter. They possessed political ability and memory—two more traits not unwelcome at a massive constructional enterprise.

In addition to these six groups of invaders, there is mentioned in some of the legends a seventh: Dravidians, from India. But such people, if they came to Britain at all, seem to have made little impression. The stories about them are few and vague.

There is curious reading in the great corpus of handed-down myth and history to do with those early colonizers. Since we cannot now separate the myth from the history we should not consider the stories as scientific evidence of anything, but we may keep them in mind, remembering that characteristics and customs of early colonizers of Ireland would probably also be those of early settlers in England. And some of those traits described in the legends would be ideal for the very real work we are going to discuss.

* Ireland was called Scotia Magna, and Scotland was Caledonia, until about the third century A.D., when, according to Bede, an Irish tribe invaded Caledonia, and its name was changed to Scotia. Duns Scotus, the thirteenth-century "Subtle Doctor," was a Scot; John Scotus Erigena, the ninth-century "Scotus the Wise," was Irish.

So much, then, for the bardic legends. Now let us see what the examiners of direct evidence, the archaeologists, say.

According to them, evolutionary forerunners of man were in England as long ago as 500,000 years, and man himself, classifiable as *Homo sapiens*, walked the hills of England soon after he appeared on earth some 50,000 years ago. At this point we must consider the question: When did animal become human?

During this century there has been a scientific debate concerning the status of early man.

One school of thought regards man as an animal, though a somewhat superior one, until about 30,000 years ago. Another school pushes the date back. Neanderthal man of 200,000 years ago and other kinds of *Homo* who lived back to 1,000,000 years ago or more, are regarded as superior enough to be closely related to *us*. Biologically, the test of species is breeding. Could *Homo sapiens* breed with *Homo neanderthalenis?* Since the latter exists only as a skeleton in the museum the definitive biological test cannot be applied. Archaeologically, the problem seems insoluble, and it becomes almost a matter of individual definition and preference.

Regardless of when animal became man, that primitive creature faced an enemy more invincible than all of the other foes that threatened—the ice. At least four times in the last two million years a giant wall of ice, hundreds of feet high, has pushed down from the north, burying the habitable valleys and plains and foothills, forcing all life in its path to migrate. Things and creatures left behind were entombed, and what the living conditions were we can hardly guess.

Geologists have long been puzzled about the cause of the Ice Ages. Many possible reasons have been proposed: decrease in the sun's energy; change in the atmospheric content of carbon dioxide, or fine mineral particles, or water vapor; local conditions; orbital variation; polar wandering; astronomic variation. Until recently the last proposed cause had fairly strong adherents. Considering the earth's orbit plane as a base, the globe's axis turns slowly, like a dying top, making a complete circuit in 26,000 years. The axis also "nods," or changes its angle of inclination to the orbit plane, in a cycle of 40,000 years. Finally, the shape of the ellipse which we describe as we revolve around the sun changes, with a period of some 92,000 years. The cumulative effect of all these changes *may* cause the earth's average temperature to vary by as much as 10 degrees Fahrenheit, which *may* be enough to cause Ice Ages. But of late that astronomic theory has lost favor. The presently popular theory is that Ice Ages were, and will be, caused by small climatic changes caused by variation in the output of radiant energy by the sun.†

We are now in an interglacial period, and have been since the last Ice Age began to melt some 18,000 years ago. But this too may pass. The great cold may come again. The wall of ice may push down again over Scotland, Scandinavia, Canada and the Great Lakes. And then Man, the Wise, unless he has become wiser than he is now, will once more have to abandon his living places just as did his ancestors before him, and move south, changing his ways of life to adapt. Some say this might not be the worst thing; it is possible that the vigorous competition for survival imposed by the conditions of an Ice Age is beneficial to those who do survive.

Since glaciations buried most evidences of the first men, nearly all of the earliest traceable artifacts and other relics pertaining to dawn men in England come from the relatively "modern" culture called Aurignacian. This culture, named for a French cave region, spread from Palestine to France. Physically, Aurignacians were generally of the flat-faced Cro-Magnon stock. Beginning about 30,000 B.C., they came in small groups across the land bridge that still joined the British Isles to the Continent, in pursuit of animals that also moved in groups—reindeer, mammoth, woolly rhinoceros. And, legend-makers note, they came to England only, *not* to Scotland or Ireland. Those lands were still largely beneath the ice. The Aurignacians were nomadic cave-dwellers. They made small flint tools and bone implements and ornaments, a few of which have been found in southern England and Wales. They may have been driven back by the last advance of the ice sheet; a Welsh cave that they had inhabited was later blocked by glacial clay.

After the Aurignacians came other rovers from the Continent, Gravettians. They belonged to a culture extant from South Russia to Spain. They also were hunters of animal herds. With them may have come the first Solutreans, from France and Spain. These people found

† I am indebted for some of this geological information to the book *The Deep and the Past* by David B. Ericson and Goesta Wollin (New York: Knopf, 1964). By analysis of cores brought up from the ocean floor they have determined that the first of the four great Ice Ages began perhaps 1,500,000 years ago; and the last, which seems to have been divided by a 40,000-year warm spell, about 120,000 years ago. They believe that the glaciations were due to the coincidence of "extraordinary topographical conditions" and fluctuations in solar radiation. During their many years of examining thousands of cores these investigators have made many discoveries—not one of them more interesting, to the layman, than this: the dominant direction of coiling of shells of a certain species of foraminifera, *Globorotalia truncatulinoides*, changes with some chemical or physical change in the water—probably the temperature.

bison, horses, and wild oxen, as well as some surviving mammoth, woolly rhinoceros and reindeer.

The number of visitors to Britain during these ages was apparently small. Indeed, from the scanty evidence available, it seems that the average "winter population" of the whole country might have been as little as 250 persons.

As the cold of the last glaciation lessened, some of the hunters settled down and began to fashion a new culture. This was the time—about 10,000 B.C.—of the Magdalenian culture on the Continent. But while Continental peoples had long before produced those marvelous cave paintings of Lascaux and the Dordogne, their British counterparts managed only some Magdalenian-style weapons and implements; remains are now found in Kent, Cheddar and Yorkshire. Perhaps England was still too cold. Or, what is more likely, the North Sea had broken through and separated the ancient Britons from the Continent.

After that North Sea separation other groups of immigrants arrived, presumably by boat, though the "sea voyage" at first was probably no longer than crossing a wide river. These were Tardenoisians, from France. Small flint-tool users who possibly brought with them Britain's first dogs, they either mingled with or chased away the islanders already settled in England. They seem to have roamed the hills in summer and lived in caves in winter. Where natural caves were not available, they dug shelters. The most puzzling thing about them was their habit of carving burins, tiny flint blades with chisel-like edges which might have been used as engraving tools.

Then came beach-loving people, the Azilians. They hunted with dogs, fished, and rarely pushed inland from the coasts. Some of them survived into the Bronze Age.

The last group of Mesolithic, or Middle Stone Age, arrivals into England were "forest folk" called Maglemoseans. They introduced "heavy industry," in the form of manufacture of stone and bone tools for use in carpentry and hunting. And they were still carrying on their trades as the climate warmed and the neolithic revolution began.

That revolution was the most significant in the whole history of early man. Before, he had been a nomadic hunter, dependent on each day's conquest to stay alive until the next. In the New Stone Age he mastered the practices of raising plants and animals, and was freed of dependence on the day. In a short time, as evolutionary time is measured, he developed improvements like irrigation and the plow and digging tools, and a hundred other things, and started civilization on its long course.

The great revolution probably started in the eastern Mediterranean (and possibly other places, like Central America, at other times), about 10,000 years ago. But diffusion of knowledge can be a slow and painful process, as all anthropologists know. Primitive tribes do not necessarily welcome radical ideas; they are quite capable of resisting an innovation even if it is demonstrably beneficial, and of putting to death the would-be innovator as a sorcerer. Significant change sometimes depends on force. In any case, it was centuries after its southern beginning that farm and village culture took hold in England. And when it did take hold, it did so in the most Mediterranean-like areas: the clement southwest coast of Ireland, and the chalk downs of southern England.

Beginning about 3000 B.C., waves of farmers crossed over the widening sea to the islands. These were the estimable Windmill Hill people. They lived a seminomadic life still, but subsisted mainly on their own flocks; big domestic animal bones are more numerous in their remains than smaller wild animal bones. Cattle breeding was their main occupation. They also kept sheep or goats, pigs, and dogs, apparently like long-legged fox terriers. And they grew wheat.

These farmers constructed large hilltop enclosures, not very accurately called "causewayed camps," such as the one on Windmill Hill near Stonehenge which gave the culture its name. These enclosures, which are Britain's oldest large structures, were made by digging circular ditches around a knoll, the ditches broken in many places by causeways, with banks behind them. Until recently it was thought that the banks were probably crowned with stockades and the entrance causeways equipped with wooden gates, forming cattle pounds or corrals which could have been used for protection—wolves were particularly dangerous then—and for cattle slaughter. That theory, however, is presently in disfavor.

They found time to engage in flint mining along with their other pursuits, and they made axes in at least one "factory" in North Wales. They were busy industrialists and traders as well as hunters and farmers.

Their relics are quite varied: arrowheads, axes and adzes for woodwork, flint blades and scrapers for leatherwork, millstones for grinding, pottery vessels patterned after earlier leather models (remember "the men of bags," the Fir Bolg?). (Fig. 2.)

Fig. 2. Tools, implements and utensils of the late British Stone Age people.

They must have brought a host of superstitions woven into a strong religion, with the significant custom of collective burial in big stone-encased tombs. Regard for the dead is considered a sure sign of cultural development; the Windmill Hill people showed great concern. Their dead were laid to rest in collective graves, or long barrows, which were covered with extensive piles of earth. Some long barrows were 50 feet wide and 300 feet long. Mostly the barrows point east-west, the general direction of the rising and setting sun. Pits with charred remains are found under the barrows, indicating some ritual preparation or sanctification of the ground. The dead were laid out one by one at the time of death until upwards of fifty individuals had been cared for. With each interment, food, tools, and occasionally pottery and flint arrowheads were placed in it before the grave was sealed. Like the Severn Valley settlers, these people built with large stones. Their long barrows were curbed with stones and boulders before the whole structures were covered with final layers of earth.

Altogether, the Windmill Hill people seem to have been a peaceful, productive folk, very important in the building up of Salisbury Plain as a focus of trade and culture.

These gentle people were the last New Stone Age arrivals in England. Next, soon after 2000 B.C., came the Beaker people, and the Bronze Age.

The Beaker people got their name from their custom of burying beakers, or pottery drinking cups, with their dead. They seem to have been well-organized, quite powerful and energetic, and possibly less peaceable than the Windmill Hill culture. Their graves contained more weapons, holding daggers and battle-axes. The Beaker people departed from the older custom of collective burial. They inhumed their dead one by one, or at most two by two, in small round graves marked by mounds. The bodies were buried knee-to-chin. Sometimes they made coffins of stone slabs, but their sepulchres are not so imposing as were those of their predecessors. Inside, however, Beaker tombs were not lacking in grandeur. They buried their great ones fully clothed, with their valuables around them—gold and amber and jet ornaments. After about 1500 B.C. the bodies were mostly cremated.

Beaker graves, or "tumuli," are so numerous that until quite recently it was a popular sport among the idle rich to dig them up, in the not entirely vain hope of finding under the earth rich Bronze Age treasure.

In life these proud warriors contented themselves with the most makeshift shelters, but in death each smallest chief had his fortress against eternity. So strong was their custom of mound burial that for a thousand years it persisted in England.

The last Bronze Age people with whom we have any concern were the Wessex people.

They appeared on Salisbury Plain soon after the Beaker people. The date must have been about 1700 B.C. Like the Beakers, they were a highly organized and industrious people, but were perhaps less belligerent. In their graves are daggers and bows as well as ornaments, but some of their ostensible weapons at closer inspection seem to have been more probably only ceremonial symbols, like our West Point and Annapolis parade swords. There is evidence that the Wessex folk were concerned less with war than with the arts and enjoyments of peace—trade and the good life. Their chiefs, that is, were so concerned; the Wessex people themselves, along with other possibly subject peoples, may have been quite sternly ruled. Their toil in mine and field seems to have made the profits which the rulers put to good use in their trading. Only the chieftains were preserved for afterlife. The ordinary folk left no trace.

Those rulers were great lords, and international financiers. Using the surplus wealth accumulated by the toilers, they bartered old necessities and new luxuries all the way from the Baltic to the Mediterranean. Among their mementoes are blue faience beads of Egypt, axes from Ireland, a Baltic amber disc bound with gold in the Cretan fashion, Scottish jet necklaces and ingenious arrow-shaft straighteners, delicate "incense cups" and tiny bowls decorated in the style of Normandy, bronze and gold and amber amulets patterned after spear-ax weapons of the North German forests, little pins from Central Europe, gold inlaid boxes, scabbard mounts, buttons. . . .

Those Wessex lords lived in busy splendor, and went into their last resting places with their martial and civil pomp around them.

Whence came this extraordinary people? The collection of international treasures does not help to locate them. It would be absurd to claim that because of the faience beads they came from Egypt, or, because of the amber and gold discs, from Crete. One must look elsewhere.

At least one archaeologist and Stonehenge authority points out many similarities which exist between the Wessex culture and the cultures of Brittany, and suggests an origin in France. Others favor Central Europe. Scottish archaeologist V. Gordon Childe has the following theory, which seems most reasonable because of its directness and simplicity:

The "Wessex people" developed *in Wessex*. As the early Beaker folk prospered, in the fashion of all island dwellers they quickly began to create their own distinctive characteristics. In a few hundred years their farming and industry had brought such wealth that a relatively intricate power structure of politicians, priests, entrepreneurs and all the other miscellany of middlemen necessary to keep their economy meshed and moving had come into being. There was probably a hierarchy ranging from famed, kinglike ruler through a sort of administrative nobility down to anonymous peasantry, all knit together in a strong commercial society, a society different enough from its Continental Beaker culture progenitor to deserve the new name of Wessex.

In any case, regardless of where they originated, those Wessex rulers, the leaders of ancient Britain, were buried amid the splendid trappings of their busy, successful, wide-ranging lives, beneath mounds which still dot the countryside today. Death and memory were matters of utmost concern to them.

All of these "people," these ancient "cultures" whose members would never have recognized themselves as such, vanished, as distinct societies, long ago, dissolved and reassociated by succeeding waves of conquest, migration, growth and decay, the endless grouping and regrouping of racial evolution.

These dawn men left but little to tell us, their descendants, of their daily ways. But they did leave lasting memorials to their gods, testimonials of fears and hopes and deep purposes—the enduring monuments of Salisbury Plain.

And the greatest of these is Stonehenge.

Chapter 3

HISTORY

As viewed today by the average tourist, Stonehenge seems to be only a cluster of giant stones. Some are standing alone, like the menhirs of other monuments, some are capped by lintels which make of them great archways, some are leaning, some are fallen. Many are missing altogether, victims of the hand of man even more than of the scythe of time. Stonehenge, the tourist thinks, is completely made of stone.

Few of the thousands of visitors to the site notice that after they have paid their shilling for admission they walk to those great stones by a path which takes them across two banks and a ditch, through a raised mound and past some marks indicating the existence of filled-in holes. Even fewer know that these non-stone parts of Stonehenge —the earthworks and holes—were for the builders and users of the structure far more valuable, in practical use, than were the imposing stones. But so it was, as shall be demonstrated in this book.

Stonehenge was so much more than a simple array of stones that its true history becomes more interesting, more marvelous, than all of the legends which have risen like fogs around it.

For this history we must thank the specialists. During the last half-century the archaeologists and anthropologists and the other experts —the diggers and daters and interpreters—have investigated the old monument with most painstaking care, and their findings have provided us with a remarkably clear report of *what* the monument consists of, and *when* it was built, and *how*. Some uncertainties still exist, but they do not blur the general picture.

The bare facts—all legends stripped away—are as follows (dating accurate to better than a century more or less):

Stonehenge was built between the years 1900 and 1600 B.C.—a thousand years or so after the pyramids of Egypt, a few hundred years

before the fall of Troy.* Its time of creation corresponded with the
flourishing of the Minoan civilization of Crete. On the Greek mainland, at Mycenae, the future conquerors of Crete had not yet reached
that state of skill which enabled them, in 1400 B.C., to build the famous Lion Gate. In Mesopotamia, Abraham was living at Harran
when Stonehenge was new; the Israelites had come into bondage in
Egypt, and had not been led forth by Moses before it was old. In
America the inhabitants had not yet felt the urge to the spectacular
that was to create the cities of Yucatan two thousand years later. And
in China, farther away than a fairy tale, men were perfecting the silk
industry and making picture language on tortoise shell to aid in the
telling of fortunes. The only other notable civilization of antiquity, in
India, has left no great stone monuments. The strange stone faces of
Easter Island are relatively recent on the Stonehenge time scale—
they were carved and erected within the last 2000 years.

The building at Stonehenge took place in three waves of activity.

First traceable construction at the site occurred about 1900 B.C.,
when the complex now called, for convenience, Stonehenge I was
started. Late Stone Age people, probably native hunters and farmers
from the Continent, dug a great circular ditch and piled up its removed earth into banks on either side. This ditch-bank circle was left
open at the northeast, to form an entrance to the enclosure, and near
that entrance, more or less on a line with the ends of the ditch, they
dug four little holes. (A in Fig. 3.) The purpose of those little holes
is not known to archaeologists, but they may have held wooden posts.
Slightly farther inward in the entrance gap, on a line with the ends of
the inner bank, the builders dug two larger holes, D and E. These
holes seem to have held upright stones. A third stone, the now-
famous "heel stone," was erected 100 feet outside of the circle,
slightly southeast of the line from the entrance. Later a narrow ditch
was dug around it and was deliberately refilled with rammed chalk
soon after it was dug. And just within the inner bank those first
builders dug the ring of 56 "Aubrey" holes.

I should emphasize here that the whole problem of determining
precise sequence of the building at Stonehenge is very difficult.
Whereas the limiting dates—1900 and 1600 B.C.—can be fixed to an

---

* The limiting dates were refined to 1900–1600 from 2000–1500 B.C. quite recently,
after I had done my first work at Stonehenge. For most of my original astronomic
calculations I postulated a building date of 1500 B.C., because that was a convenient
round number, and the most conservative estimate of age. Since the astronomic
functions involved do not change significantly in a period of 500 years I have not
reworked those calculations.

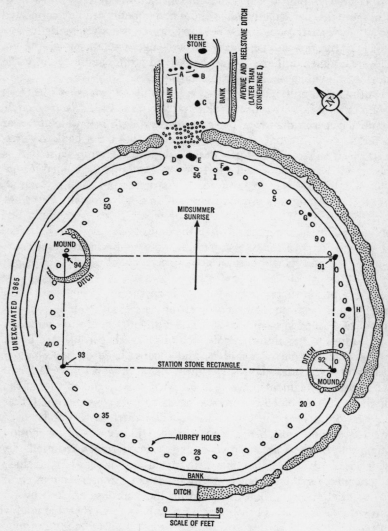

Fig. 3. A plan of Stonehenge I.

accuracy of 100 years or so, the order in which the various parts of the structure were built is sometimes impossible to deduce by archaeological methods, particularly when those components are not connected to others. Isolated holes may be impossible to date sequentially.

Thus, the first builders made Stonehenge a relatively simple enclosure, an area outlined by two banks and a ditch, entered from the northeast, with a standing stone outside.

Although simple in design, it was probably an imposing spectacle. Its outer bank, now nearly obliterated, formed a fairly true circle some 380 feet in diameter. It was an earthwork, 8 feet wide and 2 or 3 feet high. The ditch itself was just within that outer bank. As presently seen this ditch is much deeper along the eastern half; that is because it was excavated in the 1920s and only partly refilled. Originally the ditch was roughly uniform in structure all the way around, but it was extremely uneven in shape and in depth. Actually it was not a proper trench at all—it was a ring of separate pits, sometimes not connected by the breaking down of the unexcavated portions between. It was obviously a series of quarries, of no significance structurally. The pits varied in width from about 10 to about 20 feet, in depth from about 4½ to about 7 feet.

Apparently no effort was made to keep the ditch open, for soon after it had been dug it began to fill up again, with rubble that fell in or was washed in from the sides, and with whatsoever any workman had a mind to toss into it. Tools such as picks, shaped like the figure 7 and made of red deer antlers, and scoops made of the shoulder blades of oxen, meat bones (leftovers of on-the-job lunches?), and a few pottery fragments have been found at or near the bottom of the ditch, and have helped the archaeologists date its construction. Other objects found near the top of the filled-in ditch have not been so helpful, because, just beneath the surface, there can still be motion of objects in the soil. Relative dating of such things is unreliable, which is a pity, since they are easily dated absolutely. They include practically everything from prehistoric pottery through Roman coins to twentieth-century bottle tops. But they are useless as time-of-construction indicators because experience has shown that due mainly to the activities of earthworms objects dropped on loose ground may in a remarkably short time sink to considerable depths.

Beginning at the inner edge of the ditch there rose up the most impressive chalkwork of that earliest Stonehenge, the inner bank. This mound formed the rim of a circle some 320 feet in diameter, crest to crest. Glaring white, about 20 feet wide and at least 6 feet high, it must have presented an absolutely awe-inspiring barrier, at once enclosing the sacred precinct and excluding from it all unworthy or worldly things, and people. Composed of the solid chalk which makes up most of the surface region around Stonehenge, it is still quite noticeable today.

One extraordinary thing about this bank is its relative position. Practically all of the other monuments of the general Stonehenge type have their bigger encircling banks *outside* of the quarry ditches —Stonehenge, almost uniquely, has its bigger bank within the ditch. There has been much speculation concerning this puzzling exception to what was apparently a well-established rule, but at present no satisfactory explanation has been found.

The entrance which broke the two banks and the ditch on the northeast was about 35 feet wide, and its orientation was such that if a person stood in the center of the circle and looked through the entrance, he would see the sun rise on midsummer morning just to the left of the heel stone.

The heel stone, possibly the first and still one of the most controversial of the large stones which the early builders erected at Stonehenge, is about 20 feet long and about 8 feet wide by 7 feet thick. Its lower 4 feet are buried in the ground. It weighs an estimated 35 tons. The stone is a kind of natural sandstone called sarsen. Derivation of this word has never been established, but it is thought that perhaps it comes from "saracen," or "foreign," indicating the ancient belief that Stonehenge was a product of distant lands.

Actually, sarsen blocks—huge natural boulders—are found on the surface at Marlborough Downs, some 20 miles north of Stonehenge. The heel stone was very probably erected in a straight-up position, but now it leans inward toward the circle at an angle of about 30° from the perpendicular. Unlike all the other Stonehenge sarsen megaliths, thought to have been erected later, it is entirely natural in shape, bearing no marks of chipping or scraping.

Why is this coffin-shaped block called the "heel" stone? Again, the derivation is not known for certain, but it is supposed that the name was first used by John Aubrey, who said that a certain stone had a large depression shaped like a "Friar's heel." However, I have not been able to find this alleged indentation, and the celebrated Stonehenge authority R. J. C. Atkinson has told me that he thinks the mark referred to is actually on another stone altogether—sarsen stone number 14 (see end chart). The depression there, he said, somewhat resembles a right foot "considerably larger than my own."

Sometime between the 1660s, when Aubrey wrote, and 1771, the name and fame of the heel stone apparently shifted from its original stone number 14 to its present owner, because in 1771 John Smith in *Choir Gaur* placed the heel where it still, in fancy if not in fact, is today.

Thus, the heel stone has long been credited with the heel mark, and this suitable legend attached: Once there was a friar who for some reason fell out with the devil, or vice versa; the devil picked up that particular stone and threw it at the friar, struck him on the heel, and *voilà*—there you were—the friar's heel. Sometimes the word has been spelled "hele," probably because romanticists have wanted to make it appear more antique and quaint, and some less responsible philologists have wondered if the word has descended from the Greek word for sun, "helios." There is even a story that the whole stone is shaped like a heel. It isn't.

Circling the heel stone some twelve feet from its base was a ditch, presumably to indicate the stone's special sacredness.

Finally—not necessarily in time sequence, but in our listing of constructions made by those earliest Stonehenge builders—came the 56 Aubrey holes. This ring of excavations has posed a most difficult problem, if we assume that the Stonehengers had some grand design. Why were these holes so carefully spaced, and dug and then filled up? Why were there just 56 of them? 56 is not an obvious number like a multiple of the finger total, 5, or a number easy to divide, like 64. Why were there 56 Aubrey holes? I have formed a theory to account for the Aubrey holes, and I will produce this theory in Chapter 9. Meanwhile, here is a description of these most controversial things.

The Aubrey holes varied from 2½ to almost 6 feet in width and 2 to 4 feet in depth and were steep-sided and flat-bottomed. Although irregular in shape, there was little irregularity in their spacing. They formed a very accurately measured circle 288 feet in diameter, with a 16-foot interval between their center points. The greatest radial error was 19 inches, and the greatest circumferential or interval-spacing error was 21 inches. Let it be noted that such accurate spacing of 56 points around the circumference of so large a circle was no mean engineering feat.

Soon, possibly immediately after they had been dug, these holes were deliberately filled again, with a jumble of chalk rubble. Later the chalk was dug out again and refilled, often with the inclusion of cremated human bones. Sometimes the refilled holes were dug out a third time, and new cremations put in. By 1964 some 34 of the

Aubrey holes had been excavated, and of these, 25 contained cremations of humans. It was a general practice during the Stone Age to
deposit useful objects with cremations, and embedded with the bones
in the rubble were found long bone pins—for men's as well as
women's hair buns?—and pieces of chipped flint about the size of fat
cigarettes.

In 1950 a bit of wood charcoal from Aubrey hole 32 was dated
by the radiocarbon method. (Radioactive carbon 14 is constantly
produced by cosmic rays, and our atmosphere contains an "equilib-
rium" amount. It is absorbed from the atmosphere by plants, taken
in by animals when they eat the plants, and so becomes part of every
living thing. After death, the carbon 14 in a body starts a metamor-
phosis which gradually, over thousands of years, changes it into non-
radioactive, stable atoms of nitrogen, so that by measuring the amount
of radioactivity left in the body one can estimate the time of its
death.)

The age of the fragment from Aubrey hole 32 was estimated to
be 3800 ±275 years, making the date of death approximately 1850
B.C., contemporary with Stonehenge I. Not all of the cremations so
far discovered at Stonehenge have been in primary or secondary
Aubrey holes, however. In addition to the 25 excavated there, an un-
determined number—perhaps 30—have been found elsewhere, mainly
in the ditch and in the inner bank. The number of these cremations
is not known, because during the 1920s Lt. Col. William Hawley, ap-
pointed by the Society of Antiquaries to excavate at Stonehenge, dug
up many cremations and did not record their exact number, or
location.

A minor and foolish controversy has recently arisen concerning the
authenticity of some of the 55-odd cremations discovered at Stonehenge. There are skeptics who think that some of the supposedly
ancient burials are actually quite modern—the burned bones of present-day druids. Until recently the modern Order was permitted to
bury cremated remains of dead members within the Stonehenge
circle. This permission has been withdrawn, but apparently some of
the modern burials have not been located accurately in the records and
the doubters have thought that they might have been dug up and
confused with the Stone Age cremations. Such doubtings are easily
put to rest. Modern cremations create much more calcination in the
remains, and furthermore the recent druids buried only very small
packets. Whereas the average bulk of a prehistoric cremation would

about equal the size of a grapefruit, recent druid burials, says a Stonehenge curator, would fit in a matchbox.

The Aubrey holes, including cremations and all, were filled up to evenness with the surrounding ground some time after their digging. In time the grass grew over them, and they became indistinguishable from the general cover. For centuries their existence was unsuspected, until John Aubrey spotted them, some three hundred years ago. They showed as very slight depressions in the turf, possibly caused by prolonged settling of the chalk fill.

And that, apparently, was what the first Stone Age people made at Stonehenge. Stonehenge I was a ditch with two banks, three standing stones, four wooden posts, and a ring of refilled holes—the whole oriented, by alignments and approximately by an entrance way, toward the midsummer rising sun.

Was there anything—stone or hole or structure—at the all-important center of the monument? The focus of Stonehenge has never been excavated. What was or is there, if anything, is not known.

It is possible that during that first phase the builders also erected the four extraordinary "station stones," although the age of these is in considerable doubt.

As may be seen on the chart, these stones, numbered 91, 92, 93 and 94, stood approximately on the circle of Aubrey holes. They formed a rectangle perpendicular to the midsummer sunrise line of the monument. Only two of them—91 and 93—remain. These two are sarsens, very different in size and shape: 91 is a naturally shaped rough boulder about 9 feet long which now lies prone against the inner side of the old bank, and 93 is about 4 feet long and still stands upright. Its north and south sides have been slightly tooled. The other two stones, 92 and 94, are both missing. Their former presence is inferred from the nature of the holes that remain. The two missing stones stood on so-called mounds, bounded by the familiar ditch.

The ditch of 94 was roughly circular, with a diameter of some 60 feet. The ditch of 92, slightly flattened where it met the old bank of Stonehenge I, was about 40 feet in diameter, and sliced through part of Aubrey hole 19. It probably was enclosed by a low bank, as was 94, but this cannot now be verified, because Col. Hawley excavated that whole site and did not record the presence—or nonpresence—of such a bank.† At present this north mound, 94, is very hard to see.

† Oh, Col. Hawley! Although a "most devoted and conscientious excavator," in the words of Atkinson, and an efficient supervisor of the re-erection of several of the fallen stones, he dug and stripped in a fashion so "mechanical and largely uncritical," with such a "regrettable inadequacy in his methods of recording his finds and observations

A cart track and modern path by which tourists enter the enclosure have flattened its western half.

The most remarkable thing about the station stones was their rectangular placement. They were so located that each side, and the diagonal 91–93, had astronomical significance, and the diagonals intersected very close to the center of the Stonehenge I circle. The short sides of the figure lined up with the direction of the center-heel stone axis, and the long sides were almost exactly perpendicular to that axis. I believe that the station stones formed a unique figure —historically, geometrically, ritualistically, and astronomically. They were immensely significant.

During the eighteenth and nineteenth centuries there was speculation that a fifth station stone once stood just inside the bank near Aubrey hole 28, on the southwest prolongation of the midsummer sunrise line toward the midwinter sunset, and some evidence was claimed to support this theory. But later investigation has failed to produce any such evidence, and the theory must now be regarded as unsubstantiated.

When were the station stones erected? Archaeologists agree that they came after the Stonehenge I ditch-banks and Aubrey holes, because their mounds overlie those previous earthworks—but how long after? Some archaeologists think they followed very soon after Stonehenge I because they are rough, with little tooling, and thus resemble the venerable heel stone. But other authorities think they were erected much later, at the end of next wave of building, Stonehenge II. This dating sequence cannot now be determined, but I shall show later that astronomical considerations indicate an early date, and I believe these stones belonged to Stonehenge I.

The building of Stonehenge I, which began about 1900 B.C., lasted for an indeterminate number of years. Perhaps several decades were required for the various diggings and stone and wood column preparation and placement. Perhaps several more decades were spent in use of the primitive monument.

We cannot know what those earliest builders were like, nor what they felt and thought about their first handiwork, and how long they

---

and, one suspects, an insufficient appreciation of the destructive character of excavation *per se*," that he has left for subsequent investigators "a most lamentable legacy of doubt and frustration." It seems that the Colonel also had such a dislike of pottery that he may well have simply ignored and not reported objects of this nature which he may have dug up. Altogether, the Hawley excavations of 1919–1926 make "one of the more melancholy chapters in the long history of the monument."

used it. We can, however, I believe, form some idea of what they were planning and doing in those early years, by applying astronomical principles to a study of the monument considered as a whole, in space and in time. And that is why I am devoting so much attention here to a description of the objects, and the sequential timing, of Stonehenge. Familiarity with these details will be necessary for later discussion.

About 1750 B.C. the second wave of construction at Stonehenge began. This work was done, apparently, by a different race of people: the Beaker people.

These second builders brought the first assembly of megaliths, or "large stones." At least 82 bluestones, weighing up to 5 tons each, were to be set up in two concentric circles around the center of the enclosure, about 6 feet apart and about 35 feet from the center. A circle of stones was characteristic of the Beaker culture, but the ritual significance of such a structure puzzles the scholars of the past. The double circle had a small entrance on the northeast side, formed by a gap in the ring and marked by additional stones on either side of the gap. This entrance lay approximately on the line from the center to the heel stone, which was left untouched. The nearby holes B and C are hard to date, and may belong to Stonehenge I rather than to II.

The second builders also widened the old ditch-bank entrance some 25 feet by tearing down the banks and throwing the rubble into the ditch, and they extended out from that entrance a 40-foot-wide "Avenue" bordered by parallel banks and ditches. This bank-bordered roadway, now almost obliterated, originally went northeast from the Stonehenge entrance and curved right to the river Avon, some two miles away. The Avenue probably was used as a road for hauling bluestones from the river to the monument.

Now for the details of Stonehenge II—the first stone circles to appear at the site, and the broad Avenue.

The double bluestone circle, Fig. 4, seems to have been designed to form a pattern of radiating spokes of two stones each, that is to say the stones of the inner circle were matched by stones of the outer circle so that the whole resembled a short-spoked wheel. This was an unusual pattern. Could the spokes enclosing the sacred center have been meant to serve as sighting lines from or over that center? Were the stones only a ritual barrier? Or was the design a blunder? We can hardly guess, because the double circle was never completed. Some holes are missing on the west side, and two holes at the entrance were only partially dug, and stones were not placed in them. And for

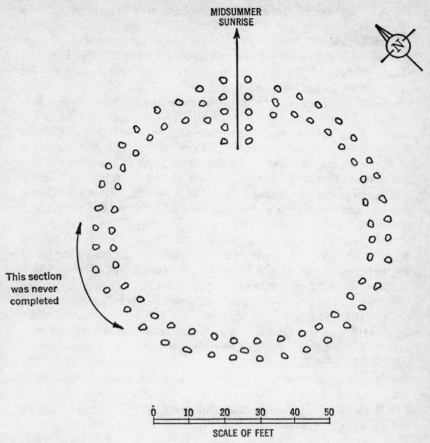

Fig. 4. A plan of Stonehenge II.

some reason the whole double bluestone circle structure was abandoned, possibly in a hurry.

How many spokes did the builders intend? The first estimate based on symmetry was 38, but in 1958 a most interesting and puzzling feature of that circle was found—a now-empty pit on its southwest side.

This pit, a large circular depression on the main axis directly across the center from the Avenue entrance, *could* have held a very large stone, possibly wide and flat-topped like a table, or altar. *Did* it ever hold such a stone—perhaps even the imaginatively named "altar stone" which now lies nearer to the center? Or was the pit always empty, intended for some other purpose? At any rate it means the intended number of spokes must have been an odd number close to 38.

The Avenue of Stonehenge II was made of two parallel banks, 47 feet apart when measured from crest to crest, with a pathway between. The ditches were shallow and the banks may have been low—the Avenue had all but vanished from sight when Dr. Stukeley rediscovered it in 1723. Recent photographs from the air have shown that this broad highway went out northeast from the entrance to Stonehenge along the midsummer sunrise line, continued about one-third of a mile into the valley, then turned east to the right, and curved on to approach and probably meet the Avon at West Amesbury. (The last few hundred yards of the route have not yet been probed.) Stukeley thought that there was a fork going toward Avebury, with the Avenue branching into north and east divisions in the valley, and archaeologists Colt Hoare (1812) and Flinders Petrie (1880) concurred. But air photography has confirmed the existence of the eastern branch only. Recent excavation has revealed that Stukeley's supposed northern branch is in fact segments of two ancient ditches which roughly paralleled each other and apparently had nothing to do with the Avenue; both seem to have been dug after the Avenue was laid down. (It is extremely hard to trace these old, long filled-in ditches. In most places the ground-bound detective is reduced to counting thistles because there are more of them where the ditches used to be. Also, other vegetable growth there is greener.) The route followed by the Avenue looks unnecessarily curved on the map, but actually those curves follow altitude contours. The route avoids steep slopes and would therefore have made the hauling of stones from the river to the building site easier.

As was the case with Stonehenge I, the building of Stonehenge II took place in a period of some 100 years or less. And as the building of Stonehenge II ended, so did the British Stone Age.

Beginning about 1700 B.C. the Bronze Age proper came to Britain, and with it the final wave of construction at Stonehenge. This date is fixed within a hundred years or so by radiocarbon dating of a deer antler found buried in the fill around stone 56.

The last builders were, apparently, the powerful, rich, commercially active Wessex people. They were excellent craftsmen who possessed quite sophisticated‡ tools and ornaments and weapons, of gold

‡ A misuse of a good word! Etymologically it means "clever, skillful, wise." Recently it has come to signify complex mechanical excellence, but purists say it should only apply to thinking organisms. People, they say, can be sophisticated; things can only be efficient, or elaborate, or ingeniously devised, and so forth.

as well as bronze. They seem to have organized themselves into groups led by warrior chieftains, but they probably preferred trading to fighting. There is strong evidence that they were in communication with the great contemporary Mediterranean civilizations of Minoan Crete, Mycenaean Greece, Egypt, and the ancestors of the traveling-trading Phoenicians. Archaeologists are traditionally conservative and ungiven to theorizing, but the indications of a Mediterranean origin for Stonehenge III are so strong that they allow themselves to wonder if some master designer might not have come all the way from that pre-Homeric but eternally wine-dark southern sea to the eternally green, pleasant and far from barbaric northern land. It is indeed a fascinating thought. Homer himself said that builders were wanderers. "Who, pray, of himself ever seeks out and bids a stranger from abroad, unless it be one of those that are masters of some public craft, a prophet, or a healer of ills, or a builder, aye, or a divine minstrel . . . for these men are bidden all over the boundless earth. . . ." (*Odyssey*, XVII, lines 282–86.)

Atkinson inclines seriously to this theory, stressing the importance of the evidence of dagger carvings and ax carvings as well as Mediterranean artifacts found in the burials of Stonehenge, and pointing out that Stonehenge is unique not only in the elegance of its construction but also in the fact that it is the *only* stone monument known to have been built by the Wessex people. Therefore it would seem not to have been part of a local building tradition, another in a continuing series, but a *rara avis*—a Minerva sprung full-grown from some father's brow without ever having had a childhood. Now how could such a complex structure, embodying very subtle, advanced concepts and even more advanced building techniques, have risen from nothing? Would there not have had to be predecessors—trial building projects? For Stonehenge, there are none—in Britain. Therefore, must it not have derived its tradition elsewhere? And therefore must not that tradition have been brought by some one man? It is an intriguing thought.

In the period labeled for convenience Stonehenge III A, the double circle of bluestones, begun in Stonehenge II and still incomplete, was taken down. The stones were laid aside somewhere—just where is not known—and replaced by 81 or more huge sarsen boulders from the same Marlborough Downs where the earlier builders had got the heel stone. These sarsens were placed in the same general area which the bluestone circles had occupied, but in a very different pattern. (Fig. 5.)

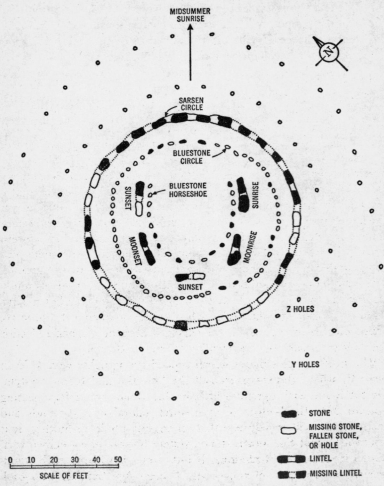

MIDSUMMER SUNRISE

SARSEN CIRCLE

BLUESTONE CIRCLE

BLUESTONE HORSESHOE

SUNSET

SUNRISE

MOONSET

MOONRISE

SUNSET

Z HOLES

Y HOLES

	STONE
	MISSING STONE, FALLEN STONE, OR HOLE
	LINTEL
	MISSING LINTEL

0 10 20 30 40 50
SCALE OF FEET

Fig. 5. A plan of Stonehenge III according to present archaeological knowledge, showing the bluestone horseshoe, the five trilithons, the bluestone circle, the sarsen circle and the approximate circles of the Y and Z holes.

First, close around the center of the monument, was erected a horseshoe of five trilithons. The word "trilithon," from the Greek words meaning "three-stone," is unique to Stonehenge, and was coined to describe a free-standing unit of two uprights capped across the top by a third crosspiece, or lintel. Second, enclosing these trilithons, was erected a single circle of 30 uprights all joined across their tops by lintels.

The horseshoe of trilithons opened to the northeast, and was so oriented that its axis corresponded to the familiar midsummer sunrise line of Stonehenge II.

This monstrous structure of new trilithon horseshoe, linteled circle and old heel stone, formed the massive stone monument whose still awe-inspiring remains so impress us today. Stonehenge III A was very nearly Stonehenge Final. The trilithons were of different height, 20, 21½ and 24 feet (including lintel), increasing in size from the northern ends to the center of the horseshoe. The central trilithon was the largest unit in the whole huge structure of Stonehenge. Its eastern stone (55) was, before it fell and broke, 25 feet long, and its western one (56) was 29 feet 8 inches long. The difference in length was compensated for by burying the western one deeper in the ground. The eastern one was embedded to a depth of only 4 feet—dangerously shallow, as the builders obviously realized, because they left a large knob on the bottom, the better, when buried, to anchor it. Stone 56, which must weigh 50 tons, is the largest at Stonehenge, and indeed is the largest prehistoric hand-worked stone in Britain.

The lintels or crosspieces which capped the uprights were held in place by what cabinetworkers call the "mortise and tenon" system. On the top of each upright a little knob or tenon was left projecting upward. Into the bottom of each lintel, placed near the ends and shaped so as to fit loosely over that tenon, was a hole, or mortise, It is noteworthy that this mortise-tenon system is more of a woodworkers' than a stoneworkers' technique. It indicates a familiarity with woodworking on the part of those early Bronze Age builders who took over the Stone Age structure. The central trilithon tenons were about 9 inches high and slightly wider than that at the base. In addition to the mortise-tenon joining, the tops of the uprights were slightly scooped out, or dished, and the bottom of the lintels correspondingly chamfered, to prevent sliding. (All of the stones erected during Stonehenge III were hand-worked, by methods which will be described in Chapter 4.)

The uprights of the trilithons were placed so close together that there was a minimum distance of less than a foot between them.

And the uprights were carefully shaped to create the visual illusion of up-and-down straightness. They were tapered—some of them in a slightly convex curve—toward the top. Such convex tapering is what architects term "entasis," and is a very advanced and sophisticated (proper use of the word!) building technique. The lintels were also shaped to create the visual illusion of vertical straightness. Their edges were widened out upward by some 6 inches, and their circumferential surfaces were curved inward slightly, the outer surfaces being somewhat more curved than the inner.

The circle of 30 sarsen stones which enclosed the horseshoe was made of smaller stones than those which were used in the horseshoe; the uprights of the circle weighed about 25 tons as compared to the 45 to 50 tons of the trilithon uprights, and the circle lintels weighed about 7 tons. The uprights were about 18 feet long, about 7 feet wide, about 3½ feet thick. They were buried to an average depth of 4 feet. Since each upright had to support the ends of two lintels, there was a tenon on each end of each upright surface, to meet the corresponding lintel mortises. And as in the case of the trilithons, these circle uprights and lintels were dished and chamfered. As a third precaution against slipping, the meeting edges of the lintels were ridged and grooved.

This sarsen circle was very carefully spaced. Its circumference was 97 feet 4 inches in diameter, and the 30 uprights were spaced uniformly with an average error of less than 4 inches. At the northeast, precisely—as might be expected—on the midsummer sunrise line, there was an entrance to this circle, made by spacing two stones (1 and 30) 12 inches farther apart than average. The center of the sarsen circle did not quite coincide with that of the old Stonehenge I circle; it was about 3 feet north of the center of the Aubrey circle. Without this displacement the sun would not rise over the heel stone in midsummer as seen through the arch 30–1. Was that displacement accidental? I think not.

The Wessex folk may have set up the notorious "slaughter stone" in the old hole E or nearby. This unhappily-named stone is about 21 feet long, and is now sunk so deep in the ground that only its upper surface shows. There may have been a deliberate attempt to bury it, by digging a pit and tossing it in. Or it may have been still standing when Inigo Jones and John Aubrey sketched it, in the seventeenth century; but one cannot be certain. Those proto-archaeologues seem to have drawn Stonehenge restored, or as they imagined it to have looked when originally built. Personally, I would not be surprised if some present archaeologist should find that this stone had been tipped out of its hole, just to the north of its outer end, a very long time ago, during the first centuries after the construction, perhaps because it interrupted the heel stone view.

In any case, the name "slaughter" for this stone is just as inapt as is the name "heel" for *that* one. Since it was originally placed upright it could not have been meant to serve as an execution block, and there is no evidence at all to indicate that it ever did, afterwards. The name was bestowed on it by recent romantics, and signifies nothing, except perhaps that Stonehenge has become such a mysterious place that everything about it tends to rouse wild and sinister thoughts. Actually, not long ago the slaughter stone proved to be most hospitable. When the dedicated digger, Col. Hawley, excavated around it he unearthed a bottle of vintage port! The vintage was circa 1801; in that year an earlier investigator, William Cunnington, had thoughtfully buried the bottle as a reward for future visitors. Unfortunately, the cork had rotted away.

The slaughter stone is a suggestive object, lying there embedded in the earth, with its visible surface rippled by light scalloping and a row of strange little holes across one end. But that scalloping was for no blood-letting purpose; it is found on many other sarsen stones (see Chapter 4); and the little holes were dug in modern times, by some enterprising person wishing to split off a piece of the huge stone. Certain beliefs, involving the druids and their sanguinary customs, seem destined to accompany Stonehenge down through more ages of obsolescence than ever it lived through as an active place of service to man.

Dating of Stonehenge III A was first done indirectly by determining the periods of use of various objects found in accompanying burial barrows (see Chapter 5). But quite recently, within the last dozen years, there has been dramatic direct confirmation for the date thus established. Atkinson detected on some of the sarsen stones more than thirty carvings representing bronze ax heads and one carving apparently of a hilted dagger of a type known to have been in use at Mycenae about the time originally estimated—1600–1500 B.C. These carvings occur on the standing stones at a height at which a workman could comfortably cut them, so they were probably put there after the stones were erected. Other evidence, mainly involving sequence of construction, helps to narrow the probable date down to 1650 B.C., plus or minus years rather than centuries.

Soon after the Wessex construction of Stonehenge III A, and possibly carrying on its design, the wave of building termed Stonehenge III B took place. In this period, twenty or more of the bluestones which had been taken down to make way for the sarsens were re-erected, in an apparently oval formation within the sarsen horseshoe. Perhaps the "altar stone" was erected. The "Y" and "Z" holes were dug. And then the bluestone oval was dismantled.

That little oval—if oval it was—is particularly difficult for the archaeologists to reconstruct because the evidence of holes and stones is scanty. The most that can be now surmised is that some kind of oval was intended, some holes were dug, some stones erected. Possibly there were lintels capping some pairs of uprights (two stones which seem to have been lintels survive), so that possibly the little formation approximated the shape and structure of the horseshoe of sarsen trilithons which enclosed it. This supposed bluestone oval may have been repented of and demolished very soon after it was begun, perhaps even before it had been completed. Another abortive attempt, like the double bluestone circle of Stonehenge II?

What was the exact shape of the bluestone oval? Or its purpose? As with the double bluestone circle, the Stonehengers mysteriously built it and then almost immediately tore it down. Another blunder? Archaeology cannot tell.

The "altar stone," as misnamed as the heel stone and the slaughter stone, also presents a difficult problem in historical reconstruction. It now lies embedded in the earth some 15 feet within the great central sarsen trilithon, but it does not lie either perpendicular or parallel to the major axis, so it may be assumed that it is not now in the place it originally occupied. The hole in which it may have stood cannot be found, however. Perhaps its hole is buried underneath it—trilithon upright 55 has fallen over *its* hole—but no such hole for the altar stone has been found. The 1801 excavator Cunnington reported that he detected a discontinuity extending 6 feet into the earth "close to the altar," but no subsequent excavators have found that possibly refilled hole either. For the moment, the where and the why of the original placement of this stone are simply not to be even guessed at. It was Inigo Jones, apparently, who first called it by its indelible present name; he might as well have called it the Plinth, or the Finger, or what-not.

Whatever purpose it served, this stone is of a material unique at Stonehenge. All of the other remaining stones there are of sarsen or of bluestone. The altar stone is of fine-grained pale green sandstone, containing so many flakes of mica that its surface, wherever freshly exposed, shows the typical mica glitter. Whereas the sarsens seem to have come from Marlborough Downs to the north of Stonehenge, and the bluestones from the Prescelly Mountains in Wales, this stone seems to have been brought from the Cosheston Beds, composed of old red sandstone, at Milford Haven on the coast of Wales, some 30 miles to the southwest of the Prescelly quarries. The stone is the largest of all the non-sarsen stones, measuring 16 by 3½ by 1¾ feet.

The "Y" and "Z" holes, dug by the workers of Stonehenge III B, were so named because originally they were considered in series with the presently named Aubrey holes which were first termed the "X" or "unknown" holes.

There are 30 Y holes and 29 Z holes. The Y's form a circle about 35 feet outside of the sarsen circle, and the Z's form a smaller circle lying from 5 to 15 feet outside of the sarsen circle.

Both Y and Z circles are irregularly spaced. The holes are generally rectangular in shape, with the long axis following the circumferences of their circles; and the depths average 3 feet for the Y holes, and 5 inches more for the Z's. There were no pressure marks on the bottoms of any of these holes which have been excavated—about half of all the holes of each circle—so that it is assumed that none of these holes ever held standing stones. Instead, they seem to have been filled again by natural processes.

The filling material of these holes has been rich in archaeologically interesting finds. At the bottom and sides of most of them the diggers have unearthed a thin layer of chalk rubble, presumably the result of a few years of weathering before deliberate filling of the holes took place. In this earliest layer there usually was also found a single bluestone fragment of the variety called "rhyolite." (For a description of the types of stone at Stonehenge see Chapter 4.) The rest of the filling of these holes was a rather uniform mass of fine brown dirt. At the bottom of this soil many natural flint pieces were found, and the rest of the fillings contained a miscellany of objects, natural and man-made: chips of both bluestones and sarsens, pottery shards from the Iron Age (500 to 0 B.C.) and thereafter, other random things on down to modern items such as pieces of tin and glass.

The Y and Z holes pose a notable number of grievous puzzles even for puzzle-heavy Stonehenge. Why was there such an unusual number, 59, of them? Why were they so irregularly spaced? Why were they never used as stone emplacements? Why is their filling material a fine soil unlike the coarse rubble of the Aubrey holes? Why was there at the bottom of practically every one of them that solitary bluestone fragment?

The archaeologists feel that they can answer the second of these questions, at least partially. Hole Z-7 was dug *after* the stones of the sarsen circle were erected, because it cuts through the filling of the ramp of the hole of sarsen stone 7; therefore we may assume that both Y and Z rings were dug after the sarsen circle was erected; therefore, it would have been difficult (but not, I think, impossible) for the builders to have spaced accurately a ring of points *outside* that existing ring of standing stones.

To the third of these questions there has been, until now, no satisfactory answer—indeed, there has hardly been even a hypothetical solution advanced.

To the fourth question posed by the Y and Z holes there have been two answers proposed. Some archaeologists believe that the fine composition of the filling material may be credited to the whim of the builders, who simply went to a different place for packing matter for the holes. Others think that these holes were not deliberately filled by the men who had dug them, or by any men—they maintain that the fineness of the material indicates that the filling of the holes was caused by the action of nature, particularly the wind blowing for centuries over a deserted Stonehenge.

To the fifth question, like the third, there has been as yet no satisfactory or even especially appealing answer proposed. Were those bluestone fragments dropped into the freshly dug holes as offerings? If so, to what power, for what purpose? Or were they symbols? Or were they nonritualistic, nonsymbolic parts of some workaday construction-gang code? What were they? No one knows.

The answers to these last four questions we may never find.

I do, however, believe that I have found the answer to the first of these questions, and I think that my answer to that question unravels the deepest riddle of the Y and Z holes; why were they dug at all? I will produce my theory later.

The final wave of building at Stonehenge, Stonehenge III C, began almost immediately after the demolition of the bluestone oval and the digging of the Y and Z holes.

In this last burst of activity—which took place probably before

PLATE 1. Aerial view from the southeast, July 1963. A round burial barrow can be seen in the foreground.

PLATE 2. The trilithons and sarsen circle, July 1963, from the north.

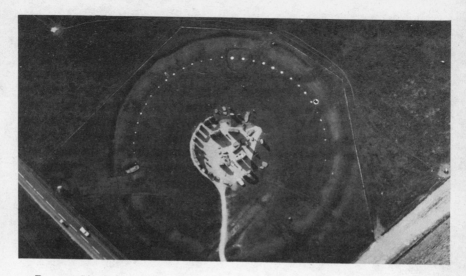

PLATE 3. Vertical survey photograph taken from a height of 1000 feet, July 1963. The ditch, Aubrey holes and slaughter stone show clearly. The heel stone is by the side of the modern road.

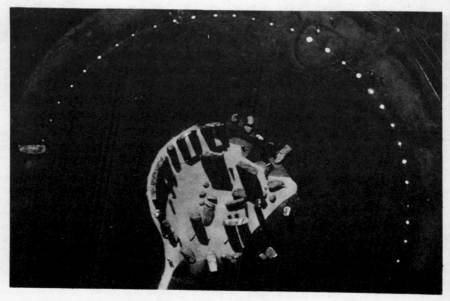

PLATE 4. The monument from a height of 500 feet, July 1963. The small, almost circular ditch near the top of the photograph is mound 92.

PLATE 5. A wartime reconnaissance photograph, August 1944. Note that trilithon 57–58 is fallen.

PLATE 6. View from the south.

PLATE 7. View from the southwest.

PLATE 8. The burial mound to the east of Stonehenge.

PLATE 9. The slaughter stone and heel stone viewed from the archway 30–1, July 1964. The camera was at a height of 5 feet 6 inches. Note that the top of the heel stone is level with the distant horizon.

PLATE 10. The heel stone from the east, showing the present tilt.

PLATE 11. A general view showing the sunrise trilithon, left of center, and the moonrise trilithon, right of center.

PLATE 12. A 1942 wartime aircraft flare illuminates the monument.

1600 B.C.—the builders re-erected the bluestones of the dismantled oval. They made the bluestone horseshoe whose remains still stand today. They also erected a circle of bluestones between the sarsen horseshoe and the sarsen circle. The altar stone may have been erected in this circle, as a towering column in line with the central trilithon.

And that was the end of the building.

The bluestone horseshoe stood a few feet inside of the sarsen horseshoe, and approximated its shape, but the smaller structure had no trilithons—the bluestones stood as monoliths. Whereas the sarsen horseshoe was made of 10 uprights, the bluestone counterpart had 19. Spacing between them was 5½ feet from center to center, and apparently the bluestones increased in height toward the closed end as did the large trilithons.

The bluestone circle, between the trilithons and the sarsen circle, had the expected opening to the northeast, but was otherwise quite irregular in shape, with spacing errors about four times as large as were found in the larger, earlier ring. This circle is now about half demolished; only 6 of its stones still stand upright, 5 more are leaning, 8 have fallen or are broken, and 10 are stumps. It is difficult to calculate how many stones it originally contained. Atkinson in 1956 thought there had been 56, 57 or 58, but four years later he revised his estimate upward, to 59, 60 or 61.

I believe for reasons which I shall give later that the figure 59 is correct. It will be remembered that this circle was made of bluestones which had previously been intended, so the archaeologists think, for the Y and Z holes—and the number of those holes totaled 59.

Whereas the stones of the little horseshoe have been tooled much more skillfully than the sarsens, with the exception of two former lintels the stones of the bluestone circle have not been hand-worked at all.

And so, with the erection of these two bluestone figures, the building at Stonehenge, which had begun some three hundred years before, came to an end. The time was about 1600 B.C., give or take 50 years.

As best we can now visualize it from outermost earthworks to center, the finished structure then consisted of the Avenue, curving up from the river; the heel stone encircled by its ditch within that Avenue; the great enclosing rings, of outer bank, ditch, and inner bank; the white Aubrey holes just within the inner bank; the four station stones, two and possibly more of them surrounded by mounds, on the Aubrey hole circumference; the Y holes, possibly empty, possibly refilled; the Z holes, likewise; the sarsen circle; the bluestone circle; the sarsen horseshoe; the bluestone horseshoe.

Stonehenge, complete, had taken about as long to build as the Gothic cathedrals which more than 2500 years later absorbed the skills and labors and love of generations of medieval men. The cathedrals were temples of worship, schoolhouses (their symbolism made clear all of the great lessons of history and morality), meeting places, memorials to faith and hope and pride.

Stonehenge may have been all of those things, and more.

Chapter 4

THE METHOD

It is dawn.

A great crowd is gathered on the plain, for it is a special occasion—the day of decision.

The sky brightens in the east. . . .

There has been laughing, earlier, and jostling to keep warm. The English night can be cool, even at midsummer. But now the people grow silent. They stand looking toward the horizon, toward the two lone trees on the skyline. Above those trees, radiating from them as a focus, the brightening sky is spreading its color out in a fan.

The priest speaks.

"People, look carefully. If God appears at the sacred place, it is good. The prophecy is fulfilled. All omens are favorable. We will build the temple here, and God will be pleased. He will protect you in life, and he will guard your spirits in death."

The chieftain, tall and strong, with the high forehead typical of his race, speaks.

"We are honored that our land has been chosen, by God himself, for his holy temple. It will be well."

The people murmur assent.

(And "yes," thinks the priest, "by this temple I will know when to call the people to this place on this one day, to see God enter his sanctuary, and by this temple I will know other things, many things." And "yes," thinks the chieftain, "this temple will be our alliance with God, a mighty fortress and monument to our power. Already we have pleased God so that he will tell the priest the good times for planting and for hunting—with this temple we will please him more—we will be great." And "yes," think the people, "a lot of work—but worth it. . . .")

The sky brightens.

The priest spreads out his arms.

Beside him, the chief stands as in prayer.

There is a moment of intolerable brilliance—an instant from eternity, the high magic moment of birth—a flash—and, exactly between those distant trees, red-gold, immense—God appears. . . .

. . . And, the next day, the enormous sanctified work began. . . .

The foregoing scene of that prehistoric village-meeting in the mid-summer dawn has all been sketched from imagination, of course, but special archaeological imagination, trained in consideration of the past, using evidence left by those people who have themselves vanished but whose traces remain to be seen.

We can shrewdly guess at their appearance and character. And we can—by reasonable interpretation of tools and toolmarks, plentiful at the site—reconstruct their work methods.

The first stage of the building seems to have been the simplest, but far from easiest. That was the simultaneous digging of the ditches and piling up of the banks.

Stonehenge is still cluttered with the instruments of this massive operation. In several stone holes and in many sections of the ditch were found the old picks and shovels. Col. Hawley dug up eighty picks in the part of the ditch he excavated.

The picks are antlers of the red deer; the shovels are shoulder bones of oxen. There may have been other bone tools—some bone fragments resemble modern rakes—and there may have been stone tools other than flint chips, and wooden tools, which have rotted away. There probably were vegetable or leather baskets of some sort, also now gone back to earth.

Marks on the sides of the ditch and holes show that the picks weren't just jabbed in by hand. Chalk is too hard for that; picks would wear out almost as soon as patience. The antler tines were probably scraped to some sharpness and driven into the chalk by pounding, then prised sideways to loosen chunks.

The resulting rubble was doubtless shoveled into the baskets and taken—by the original loader or by a human chain—to the dumping spot.

Copies of those old tools have been made and workmen of average strength and skill set to wielding them, as a test of efficiency. It was found that a man can excavate a cubic yard of chalk as solid as that of Salisbury Plain in a nine-hour day. Surprisingly enough, even with the best modern picks and shovels a worker apparently cannot do much better; the tests showed that with the modern tools the cubic yard could be dug in seven hours compared to the Stonehenger's nine.

Each digger probably needed two helpers to fetch and carry the baskets of chalk. Since the bank contained 3500 cubic yards, this work must have required 35 days for 100 diggers, with 200 extra helpers. Allowing for "days off," and days when rain made the chalk too slippery to work, the building of the bank probably required no more than one summer season for a few hundred men to complete.

Placement of the stones of Stonehenge I, II and III required more, and much more elaborate, effort than the digging and piling up of the chalk.

The ordinary tourist, standing by those huge silent stones which look as though they had been there since time began, does not recover enough from his somewhat overawed general impression of mysterious antiquity to think of asking such a simple, practical question as "How did those stones get there?" He would almost rather ask how the redwoods of California grew, or how Niagara Falls was born. For him, the medieval belief in Merlin's magic nearly suffices; who could even begin to guess how such an elemental creation as the great stone temple was ever called into being?

And yet the archaeologists—those ingenious diggers into the minds as well as the mounds of the men of the past—*have* asked that question. And they have answered it, with commendable imagination where diligent investigation has not provided clues. They have reconstructed where they could, supposed where they had to, and pieced together a very reasonable and convincing theory as to where the stones were found (not in Ireland!), how they were shaped and dressed, how transported, how erected.

It presents a rather startling picture. Instead of the traditionally described primitive savages incapable of "culture" beyond that required to daub themselves blue with woad, it is now becoming apparent that those Stone and Bronze Age Britons were highly organized, technically skillful, manually dextrous, mentally subtle folk. The story of the "how" of Stonehenge is every bit as interesting as its "when" and "what."

Let us start with the bluestones (the so-called bluestones, we should say—because the word as used at Stonehenge applies to five separate kinds of rock which have in common only a bluish tint, best seen when wet, and an igneous origin).

Most of the bluestones are dolerite, a coarse-grained greenish-blue stone, but twelve of these stones are now buried stumps of interesting composition: five are volcanic lava, darkish gray-blue in color, called rhyolite; four are a type of darkish olive-green volcanic ash; two are a gray-blue Cosheston sandstone, and one is a bluish calcareous ash. Geologists find much to speculate on in the varying natures and placements and weatherings of these different types of stones, but for the nonspecialist the most interesting fact about the various blue-stones is this: all three main types—dolerite, rhyolite and volcanic ash—occur naturally close together in a small area about a mile square in the Prescelly Mountains of Wales—and *only* there.* "There can thus be no doubt now," notes Atkinson, "that it was from this very restricted region that the bluestones were chosen and brought to Stonehenge." That distance, as the crow flies, is 130 miles—as the rollers roll, the raft floats, and the rollers roll again (see Fig. 6), the

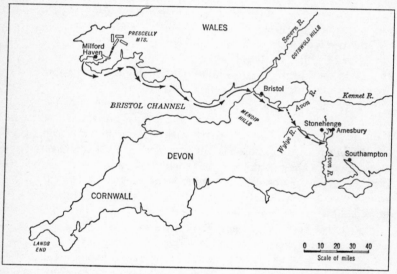

Fig. 6. The probable route of the bluestones from the Prescelly Mountains in Wales to Stonehenge.

distance is 240 miles. Bearing in mind that those eighty or more bluestones weighed up to five tons each, that is quite a long way. Nothing like this astonishing feat of transportation was ever attempted by any other people anywhere else in prehistoric Europe. The only comparable performance, indeed, was the moving of the other big stones, the sarsens, to Stonehenge.

As the map shows, the probable route began at the bluestone

* Full credit for the discovery of this source of the bluestones goes to Herbert Thomas of the British Geological Survey.

source in the Prescelly Mountains, went southwest to the sea at Milford Haven, followed the coast all the way to Avonmouth, then went up the Bristol Avon and Frome rivers, overland to the river Wylye, down that stream to the Salisbury Avon, and up that river to Amesbury and the Stonehenge Avenue. Total overland distance: about 25 miles. Total water distance: about 215 miles. This route seems most probable because it makes maximum use of safe waterways. Furthermore, there is circumstantial evidence: near Milford Haven occur the only two kinds of bluestones not found in the Prescellys—Cosheston sandstone and calcareous ash. Presumably the Stonehengers picked up these stones on the way. Also, in a long barrow near that part of the river Wylye which is supposed to have served as a watercourse for the route there was found a piece of dolerite.

It is probable that the builders mapped routes by water as much as possible, because it is much easier to move stones over water. Their land progress was doubtless made in not the hardest way—surely they must have used all their considerable skills, and eased the boulders along on sledges which rolled over an endless belt-tread of tree trunks, the sledges being pulled by teams of men using ropes of twisted hide.

It seems a brutally laborious method. However, it must have been efficient. The Stonehengers apparently did not know of the wheel, but perhaps would not have used it if they had possessed it. The Egyptians had the wheel centuries before Stonehenge was built; yet, even so, they were using a sledge-roller method for hauling pyramid stones as late as 500 B.C.

In 1954 the British Broadcasting Corporation televised a program re-enacting the drama of the Stonehenge stone-moving as imagined by the archaeologists. Teams of men and boys went through all the motions.

They took a replica of a bluestone, made out of concrete, and lashed it to a simple wooden sledge. Then they hauled. It was found that 32 sturdy young men could just pull a 3500-pound load up a 1-in-15 (4°) slope. When rollers were placed under the sledge, in the house-mover's technique of continually taking the rollers from back to front as they come out from under, the manpower necessary to haul that load was reduced to 24. Thus, the experimenters decided, about 16 men per ton would be sufficient to move stones, by this means, a mile or less a day.

To recreate the probable method of transport over water, the experimenters made three "canoes" of wood, latticed them with four crosspieces, and loaded onto this pontoon-raft the bluestone replica. The raft then drew 9 inches of water, and a crew of four punted it along easily. Indeed, in quiet water, one boy could have handled it. What would happen if such a raft went into water deeper than its punt-pole was not tested on this BBC program, but the supposition is that crude sails and oars would suffice to propel and guide it.

There seems a possibility that some of the bluestones—not the dolerites—were brought to the Stonehenge area hundreds of years before that monument was begun, and placed in a structure about a mile to the northwest.

The larger sarsen stones apparently came to Stonehenge from a source much closer than Wales. It is almost certain that those eighty huge blocks were brought from the Marlborough Downs, only 20 miles to the north. There such blocks were at that time to be found lying on the ground, presenting no problem of quarrying. As John Wood wrote in 1747, "Marlborough Downs, or rather Duns or Dunes, are covered with vast quantities of stones of the very same kind with the light coloured pillars of Stonehenge . . . scattered upon the surface of the earth . . . vulgarly called the Grey Weathers. . . ."

The sarsen route began at or near Avebury, and it may have been an important ritual to sanctify the stones as they were pulled through that monument's sacred circles on their way to their final, holiest use at Stonehenge, as pilgrims going from a parish church to a cathedral might be blessed. Some of the stones may have been used as part of the Avebury structure before they were moved south.

The sarsens average 30 tons, with the largest, the trilithon uprights, weighing up to 20 tons more. At the rate of 16 men per ton, it must have taken 800 men to pull such stones, with perhaps 200 more needed to move the rollers, clear the brush, guide the sledge and so on. The task of moving the sarsens from Avebury to Stonehenge would have kept a thousand haulers busy for seven full years.

In 1961 Patrick Hill, geologist of Carleton University, Ottawa, proposed an alternate to the generally accepted route. He put forward the theory that the sarsens were found in an outcrop south of the Kennet River (see Fig. 7) and were hauled south to the Avon River.

Fig. 7. Two routes suggested for the sarsen blocks. The first from Marlborough Downs through Avebury and over a ridge, the second along the course of the river Avon.

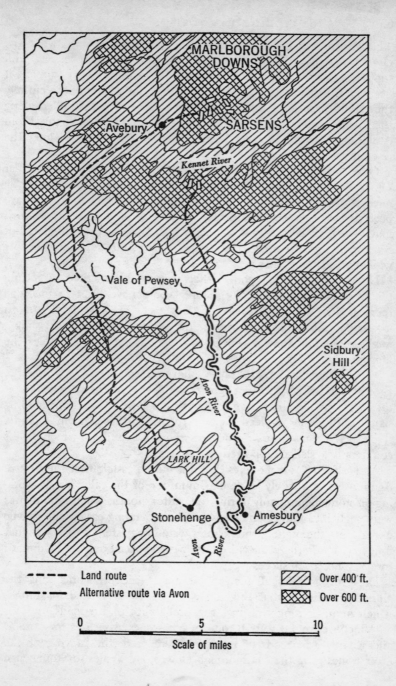

MARLBOROUGH DOWNS

SARSENS

Avebury

Kennet River

Vale of Pewsey

Sidbury
Hill

Avon River

LARK HILL

Stonehenge • Amesbury

Avon River

- - - - Land route
- · - · Alternative route via Avon

▨ Over 400 ft.
▧ Over 600 ft.

0   5   10

Scale of miles

That stream is only two feet deep now, but in those days, when the climate may have been different, it may have been deeper. Or, Hill supposes, the people may have dammed it near Amesbury to make it deeper. In either case, even if the stones weren't water-borne entirely they could have been partially buoyed up as they were pulled along the stream bed, or they could have been sledged along the bank. From Amesbury they could have been hauled to Stonehenge along the Avenue.

Other interesting facets of the Hill theory are his speculations that (A) the stones were slid down the 150-foot drop from the hilly ridge north of the Vale of Pewsey into that valley; that (B) they may have been sledged over ice or hard-packed snow, or both, in winters which may have been colder; that (C) they may have been accumulated at staging points and moved by different gangs, at different, possibly widely separated, times.

Discussing (A), he theorizes that a reasonable method of getting the huge blocks down a steep slope would have been to unsledge them at the top and let them slide down on tracks made of smooth logs laid end to end parallel to the direction of motion. Such sliding might score the stone, he feels, and cites as possible examples the long grooves in sarsen No. 16.

With regard to (B), Hill says we have no way of knowing how the Bronze Age British winters compare with those of the present, but *if* they were colder, the ice and snow would have made stone-hauling significantly easier. Indeed, he says, on smooth ice, down a gentle slope such as this route provides for 17 of its 21 miles, 25 men or less could pull a sledge-supported 50-ton stone.

And as for (C), he believes that it is quite possible that the Bronze Age movers very wisely made maximum use of the calendar, by working on stone-hauling only during the winter months when farm chores were negligible and ice and snow made the hauling easier, and maybe only during the nights, when chores were nonexistent and it was that much colder. Perhaps, he thinks, they even spaced out the hauling of some stones over successive winters.

However, as we shall see later, there is evidence concerning the climate, indicating that England was then in a thermal maximum, which argues strongly against the probability of ice roads.

Whichever routes were taken, for bluestones as well as for sarsens, and whatever methods of transport were used, the moving of the great stones from the Marlborough Downs and Wales to Stonehenge must have been a major undertaking for a good part of the population of southern England.

After the stones had been brought to the building site, how were they shaped, dressed, polished, erected?

In this phase of the reconstruction the archaeologists have more facts to go on. There are these bits of evidence to guide them in their guessing: a few stone chips and a lot of tools, and considerable knowledge of the techniques used by contemporary craftsmen in other parts of that pre-writing but not pre-exchange-of-information world.

Doubtless some rough shaping of the stones was done at the source —Wales for the bluestones, Marlborough Downs for the sarsens. Boulders larger than desired would almost certainly have been split to approximately the required size before hauling began. This splitting could have been done by wedging into cracks, sometimes the wedges being soaked with water to swell them, or by direct pounding.

There might have been used a comparatively advanced technique of hot-cold-bash. In this method a desired line of cleavage is laid out, fires are lit along this line, then cold water is poured suddenly on the heated surface. While the area is in hot-to-cold stress, it is bashed by mauls or heavy stones, and a chunk of the stone may break off, or the line open into a crack.

When the natural or rough-hewn boulders reached Stonehenge, more delicate shaping and polishing was administered. This was done in several ways—none of them quick, none of them easy.

Probably most of the shaping of the stones was done by direct bashing, with large mauls weighing as much as 60 pounds. These mauls were naturally-shaped boulders, conveniently found lying around. Since sarsen stone is very hard, the mauls were of the same sarsen material.

Maul-pounding wore away the surface surely, but very slowly. Modern experiments have shown that a strong man bashing at a sarsen with a maul can knock off 6 cubic inches per hour. Atkinson figures that at the very least 3,000,000 cubic inches of stone were removed from the Stonehenge sarsens. That task must have cost nearly 500,000 man-hours of labor.

After such pounding had reduced the stone to a very rough approximation of the desired final shape, slightly more delicate dressing methods were employed.

By skillfully directed application of the mauls, long shallow grooves, about 2 or 3 inches deep and 9 inches wide, were hollowed out. These usually ran the length of the stones. Then the ridges between these grooves were bashed down. This sideways bashing with the heaviest mauls produced some chips, about the only chips resulting from any of the stages of dressing the stones.

This coarse dressing was sometimes followed by finer, more precise shaping. Little grooves no longer than 9 inches or less, 2 inches wide, and ¼ inch deep, were scored. Sometimes several deep short grooves were dug, perhaps to remove unsightly bumps.

Sometimes, by no means always, the whole grooved-ridged surface was pounded into uniform flatness by the original simple mauling technique.

And as a final grace note, occasionally the surfaces thus leveled were further smoothed by grinding. Heavy sarsen stones were pulled back and forth over them, perhaps with crushed flint mixed in water used as an abrasive.

By one or another or several of these means any stone could be shaped and smoothed to a remarkable degree of polish. Even the mortises and tenons could thus be made to fit together quite accurately.

The Stonehenge "carvings" discovered in 1953 were doubtless produced by methods similar to the coarser dressing techniques. Sarsen stone cannot be cut by flint, and even bronze cuts it only with the greatest difficulty, so it is probable that the prehistoric axes and dagger were wrought by "delicate" pounding and scraping. (Modern initialing of the old stones, "in search," as the archaeologists bitterly say, "of squalid immortality," has been done by modern, stone-mason, methods. Most curious of these modern inscriptions are the question mark shape, with "LV" in the loop, dug into stone 156 about 130 years ago, and the very-visible IOH:LVD:DΣFΣRRΣ on stone 53. Because of that Greek Σ this inscription has been thought classic, but actually it was incised in the seventeenth century for [by?] somebody named Johannes Ludovicus [John Louis] DeFerre.)

Only a few of the bluestones at Stonehenge were dressed, but all of the sarsens of Stonehenge III show signs of some dressing. And in most of the cases where the dressing was uneven or incomplete, the stones were so placed that the smoothest side was inward, the better to be seen by those standing within the sacred circle.

Today, many of the sarsens look hopelessly rough and pitted, as though they could never have been shaped by tools. But that is because of some subsequent defacement by time, the long millennia of weathering. And the weathering has not been uniform. Sarsen is not homogeneous, and wind and water have hollowed out deep holes.

Of all the megalithic monuments in Europe, Stonehenge has the most extensively dressed stones. The nearest competitors are chambered tombs in New Grange, Ireland, and Maes Howe in the Orkney Islands—whence came to King Arthur's Table Round the magical Gawain and his brothers Aggravaine, Gareth and Gaheris.

To erect the stones, the builders first dug the holes, their depths corresponding to the length of stone to be buried, their lateral dimensions about a foot greater than those of the stones. Three sides of the hole were made vertical but the fourth sloped at a 45° angle, to form a reception ramp. When a stone was ready for placement, the side of the hole opposite the ramp was lined with thick wooden stakes to keep it from being gouged by the end of the descending stone. The stone was rolled over the hole and tipped into the ramp, its end sliding harmlessly down the stakes. Then, with the aid of whatever hide or vegetable-fiber ropes and tackle they could think of and make, some 200 men could heave a 30-ton stone upright. And as soon as it was vertical, all empty space around its foot was filled, in an understandably frantic hurry. Anything and everything the laborers could reach they threw into the gaps, to keep the stone from falling over: mauls and other tools, rocks, bones, scraps, turf—everything went in. The packing was then tamped hard. And then, probably, the standing monster was allowed to rest there for many months, so that its packing could harden and all settling cease. It was of course aesthetically important that the tops of the sarsen trilithon and circle uprights be level, which must have meant more measuring, bashing and shaping, after the settling.

It is significant to note that the bottoms of the uprights were carefully cut down to dull points, so that after they were packed into their holes the stones could still be adjusted slightly by turning.

How the final feat of construction—the placing of the massive lintels across the tops of the uprights—was done, we can only guess. There are no records or artifacts or other evidences to help.

Assuming the ingenuity indicated by other stages of the work, and the demonstrated toolmaking ability and organization of men of the time, it seems that the method may well have been to rock the lintels up on a rising latticework tower of logs. That is, the lintel was put on the ground next to the bases of the two uprights it was to cap, then logs were laid perpendicular to it, touching it. Then it was rocked over onto this layer of logs. Then the platform of logs was extended out to cover the place the stone had been, and raised by two more layers, parallel and perpendicular. The stone was then rocked back onto this higher layer. The original layer was then raised by two courses and the stone rocked back, and so on, until the wooden tower top was level with the tops of the uprights. The last step was to roll the stone over so that its mortise holes dropped onto the projecting tenons of the uprights.

Such a latticed tower would require about a mile of 6-inch diameter logs cut into 20-foot lengths with notches similar to those in a log-cabin wall.

Another method of erecting the lintels could have been to haul them up an earthen ramp as was done with stones for the pyramids. This method was suggested by S. Wallis in 1730, and as late as 1924 Edward Stone, in his authoritative book *The Stones of Stonehenge*, put forward the same theory, concluding that the lintels could have been pulled up ramps sloping as steeply as 40°. The work involved in piling up and removing such a ramp for every one of the 35 lintels at Stonehenge would have been prodigious—an earth-moving task far greater than the entire ditch-bank operation of Stonehenge I—and since recent investigation conducted in the surrounding area has failed to produce any evidence that such earth ramps were ever dug, it is now thought that this method was not used.

Wooden ramps could have been used, but they would have required much more timber than wooden towers, and would have been much more dangerous.

The curious may wish to compare construction of Europe's unique stone monument with that of the other outstanding antique stone structure, Egypt's pyramid of Cheops. This, called the great pyramid, was made of 2,300,000 blocks of stone averaging 2½ tons, the heaviest 15 tons. It was 481 feet high, and covered 13 acres. Like most of the other eighty-odd major pyramids it was on the west, or "death," side of the Nile; it was oriented true north-south-east-west, with a maximum error (on the east side) of ⅒ of a degree; it was erected in a few years by tens of thousands, perhaps hundreds of thousands, of men more slave than free.

Stonehenge, less massive but quite as cunningly contrived, was built over a period of three centuries by hundreds, or at the most thousands, of workers. The status of those workers we cannot know. But we can shrewdly surmise that their attitude toward their task was very different from that of the Egyptian laborers. The great pyramid was certainly one man's tomb—Stonehenge must have belonged to everyman.†

For generations the work on Salisbury Plain must have absorbed most of the energies—physical, mental, spiritual—and most of the material resources of a whole people.

The total work estimate for Stonehenge I, II, and III is as follows:

MINIMUM WORK TOTAL IN MAN-DAYS

Digging ditch, making bank: 3500 cubic yards, at 1 yard per man-day	3,500
Carrying for above	7,000
Digging 5000 cubic yards for Avenue banks, leveling, survey, etc.	6,000
Carrying for above	12,000
Transporting 80 bluestones, average weight 4 tons, 24 miles by land at 100 men per stone, 1 mile per day	192,000
216 miles by water at 10 men per stone, 10 miles per day	17,280
Erecting Stonehenge II at 20 man-days per stone	1,600
Transporting 80 sarsens, average weight 30 tons, 20 miles by land at 700 men per stone, 1 mile per day	1,120,000
Dressing, shaping sarsens: 3,000,000 cubic inches of rock powder at 50 cubic inches per man-day	60,000
Cutting with stone axes, hauling 300 logs for lattice tower, 2000 rollers, at 1 man-day per log	2,300
Making 60,000 yards of hide rope at 1 man-day per yard	60,000
Erecting Stonehenge III at 200 man-days per stone	16,000
TOTAL MAN-DAYS	1,497,680

To that staggering total of 1,500,000 man-days of physical labor must be added an incalculable but certainly large amount of brainwork. The organization, administration and logistics—all of the "man-handling," if one may so use the expression, necessary for such a vast communal operation—must have been complex and difficult in the extreme. Each worker had to be fed and clothed during the operation, and men, or women, would have been needed to keep the supply lines filled. And the actual planning and engineering were, as we shall see, extraordinarily elaborate and of the highest degree of excellence

† On some pyramids' stones there may still be seen the names of work gangs daubed in red ocher—"Vigorous Gang" and "Enduring Gang" and "North Gang" and so forth. How interesting it would be if some day an investigator should find similar notations of British work gangs at Stonehenge!

then, in Britain, possible. All of this "desk work" must have required the continuing contribution of many men, the best and brightest in the land, for many generations.

To what modern effort may one compare the building of Stonehenge? May one liken it to the present U. S. Space Program? That comparison may not be so wide of the mark.

The Space Program now absorbs, directly or indirectly, the energies of about 1 person in 1000 of our United States employed population. Stonehenge must have absorbed at least that proportion of the national energy—England's total population then was apparently less than 300,000, of whom doubtless 1000 worked on the monument.

The Space Program takes about one per cent of our gross national product; Stonehenge must have taken at least a corresponding amount. Their building effort may have required more of them than our Space Program does of us; correspondingly, it could have meant much more to them.

Chapter 5

OTHERS

Stonehenge was not alone.

It was a unique structure, but it was surrounded by activity. Close by were many other sites which may have been contemporary or older, and which were possibly related to it culturally. These sites, in roughly estimated chronological sequence, were the long barrows, the Cursus, Woodhenge, the Sanctuary, Durrington Walls, Avebury, the round barrows, and the monstrous and mysterious Silbury Hill. (Fig. 8.)

The barrows are mounds containing burials; nearly 350 of them have been found within a few miles of Stonehenge—more than in any other region of equal area in Britain. Authorities feel that they may have been placed around the monument as present-day graveyards surround churches, indicating the religious nature of the structure.

The oldest of these mounds are the "long barrows," long mounds made of chalk rubble dug from flanking ditches, with the actual burials, containing many bodies, at one end. These were built by the Windmill Hill people between about 3000 and 2000 B.C. The farming and cattle-raising of these people contributed very importantly to Stonehenge, making possible that freedom in time and environment without which such structures could not have been conceived and erected.

The most remarkable Windmill Hill long barrow yet discovered is at West Kennet, some 16 miles north of Stonehenge. This tremendous earth and stone sepulchre, 350 feet long and tapering in width from about 75 feet on the east end to about 50 feet on the west, is the largest known prehistoric tomb in England and Wales. Constructed well before 2000 B.C. and in use for at least three centuries, it demonstrates that the Stonehenge region was regarded as of

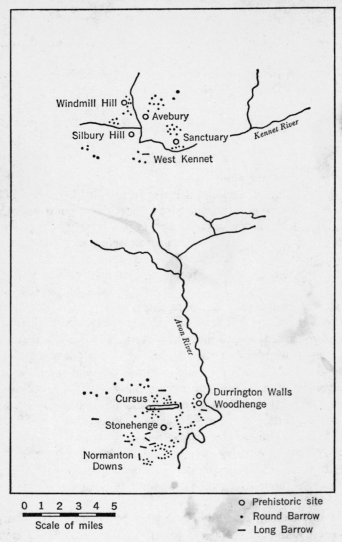

Fig. 8. A map of other prehistoric constructions in the vicinity of Stonehenge, giving only a suggestive representation of the long barrows and round barrows, which are numerous and scattered throughout the areas around Stonehenge and Avebury.

supreme religious importance long before actual construction of that monument itself began.

The West Kennet long barrow also demonstrates building ability of a very high order. It is regarded as one of Europe's outstanding megalithic structures.

Originally, the mound was made of chalk rubble, excavated from two ditches some 60 feet from each side and thrown over sarsen stones. A row of sarsens edged the sides and back. At the front, or east, end was the burial place itself, a central corridor from which five tomb-chambers branched, with two at each side and one at the end. When excavated some ten years ago the five chambers were found to contain bones of more than 40 persons, including perhaps ten children. The bones were on the floor and had apparently been entombed at different times. Indeed, it appeared that earlier inhabitants had been brusquely swept aside to make space for later arrivals. Many bones and skulls had been removed. Some pottery vessels were found with the bones.

The chambered sepulchre was about 35 feet wide, 43 feet long, and a maximum of 8 feet high internally. It was entered through a wall of large standing stones which deepened into a semicircle at the center.

This tomb was closed, for what reason we do not know, in a very thoroughgoing way. The five burial chambers, with bones and all, were crammed full of chalk rubble, broken pottery and various other material, including some animal bones. Then the central passageway was similarly filled. Finally, the semicircle around the entrance was partly filled with boulders and blocked by three enormous stones. The largest of the West Kennet stones weighed about 20 tons.

For rituals concerned with death, one can hardly imagine a more impressive place than this vast earthwork, flanked by white trenches, stone-edged, with more great stones guarding the gateway to the tomb.

After the long barrows for communal burial came the individual-burial round barrows of the Beaker people and their successors.

These barrows were of three types: "bowl," "bell," and "disc." The bowl barrows were simple round hummocks, occasionally ringed by a ditch. They were most numerous around Stonehenge. The bell barrows had larger mounds, with flat areas between mound and ditch, and perhaps with an outer bank. Most of them contained bones of men. The disc barrows were the most modest of these individual burying places. They were just flat areas with a small elevation now almost invisible in the center and a ditch and bank outside. They seem to have been exclusively the tombs of women.

Excavation of long barrows has produced few artifacts, but the individual round barrows have yielded objects of extreme interest. Weapons have been found with both cremations and skeletons, particularly bronze daggers of a type common in Brittany. Some of these daggers had handles made bright by the insertion of many tiny gold pins. Ornaments have also been found: bronze pins, perhaps from Bohemia; blue faience beads from Egypt, usually with the women's remains; amber beads from Central Europe, jet beads from eastern England. The burial barrows around Stonehenge, doubtless containing the bones of many of the builders, show that the site was not only of sacred significance. Salisbury Plain was a meeting place for far-traveling warriors and traders as well as worshipers.

Next in our list of interesting sites near Stonehenge is that extraordinary earthwork called the Cursus. Its name is from the Latin word meaning "course," and was given to it by the druidophilic Dr. Stukeley, and is as apt as any other name which might now be applied. We know absolutely nothing about the purpose for which it was made. We can only assume, as did Dr. Stukeley, that it served as some sort of a ceremonial path or enclosure. It is an area which was apparently never a raised mound, some 100 yards wide and 1¾ miles long, bordered on either side by a low bank and ditch similar to those which bound the Avenue. It lies about a half-mile to the north of Stonehenge and runs almost due east and west. Its east end terminates a few yards from a north-south oriented long barrow, and its west end, which widens out to about 145 yards, encloses two round barrows. It seems to have been built about the time of the building of the Stonehenge Avenue.

Woodhenge, as its name implies, is—or was—a sort of Stonehenge in timber. And because it was made of timber it defied time so ineffectively that it has all but vanished. It was only discovered, by air photography, in 1925.

Lying about two miles northeast of Stonehenge, it was originally a circular area some 200 feet in diameter enclosed by an outer bank and an inner ditch, steep and flat-bottomed, containing six concentric rings of holes, the smaller rings being slightly oval, the innermost ring lying some ten feet from the center of the enclosure. We know the holes held wooden posts because many rotted stumps survive. But we do not know what the posts supported, if anything. The most likely present supposition is that they were the supports for a roofed structure, high in the middle and slanting down toward its inner and outer edges, probably open at the center, like a doughnut. Archaeologists think that the inner Woodhenge structure was erected after the outlying ditch and bank were placed, probably by the same Secondary Neolithic people who started Stonehenge.

About five feet southwest of the center of Woodhenge, diggers have found a grave containing one of the very few bits of evidence that human sacrifice might have been practiced in prehistoric Britain —the skull of a child, split open before burial. The child was about three years old. Archaeologists assume that there was almost certainly ritual murder here, but the evidence is not conclusive. South of the child's grave, 45 and 60 feet from the center, were two holes intended for upright stones.

It is tempting to suppose that Woodhenge was the living place for the workers who were building Stonehenge, a sort of general barracks. But there is little archaeological evidence to support this theory. Few artifacts have been unearthed at the site, and those few have been broken bits of pottery and other such odds and ends, not the kind of household refuse which residents of even the most impersonal sort of barracks might be expected to discard. The present theory is that Woodhenge, like Stonehenge, was probably a temple or public meeting hall or both. Possibly it was a predecessor, an early effort which became a crude model for its southern neighbor. And possible there was some astronomic significance to Woodhenge. The long axis of the smaller ring ovals points approximately toward midsummer sunrise, as does the Stonehenge axis, although the center of the Woodhenge entrance, while in the northeast like the Stonehenge entrance, lies slightly to the north of that sunrise line. Mrs. B. H. Cunnington, who with her husband excavated at the site in 1926–1928, thought that a flat sarsen stone—called the "cuckoo stone"—a quarter of a mile away from the structure was part of the Woodhenge complex because it lay almost exactly due west of the center, and there "would have been a clear view [of it] . . . between the uprights from the center."

On Overton Hill, some 17 miles north of Woodhenge, is another large prehistoric wooden building site, "the Sanctuary." Like Woodhenge, this site included six rings of holes for wooden posts; unlike Woodhenge, where all the posts seem to have finally held a single structure, the Sanctuary's six rings seem to have been dug at quite different times and to have held posts which supported possibly three different successive edifices. And unlike Woodhenge, where apparently only two stones appeared, the Sanctuary seems to have ended as two circles of standing stones, with no wooden structure at all.

The Sanctuary enclosure was apparently unmarked by ditch or bank. When it was finally completed, it was about 135 feet in diameter, but it seems to have begun as a small ring of eight posts about 8 feet from a center post. It may be that this simple structure was no more than a ritual open circle, with the posts possibly carved and/or painted in the fashion practiced by Indians of Virginia as late as the sixteenth century. They used such circles of decorated posts as markers to dance around. The purpose of the Sanctuary center post is unknown. It may have been only a reference point for later construction.

Some time after the first circle of posts had been erected a second wave of activity took place. Two more rings of postholes were dug, about 2 and 7 feet outside of the original ring. These posts may have held a roof which protected the inner circle. And these posts served for many years. Most of them stood so long that because of rot or other reasons they had to be replaced.

Then came a third burst of construction. The original center post and the two phase II rings of posts seem to have disappeared. The original inner circle of eight small holes gave way to a new ring of six bigger holes, and two new outside hole-rings, about 45 and 65 feet in diameter, were added. It is thought that these three rings of posts may have supported a circular roof, possibly open in the middle, which sloped upward toward the center.

All of this construction in wood was apparently done by the same Secondary Neolithic people responsible for Woodhenge and for Stonehenge I.

Finally, the Sanctuary was turned from wood to stone. New builders, possibly the earliest Beaker people, took down the wooden posts and whatever structure they supported and made of the site another, much simpler, Stonehenge—two concentric circles of standing sarsen stones. The inner circle was about 45 feet in diameter and the outer was about 135 feet across. At the same time these builders seemed to have erected two rows of sarsen stones to border a 50-foot-wide avenue going from the Sanctuary all the way to Avebury, a mile and a half to the northwest.

The purpose of the Sanctuary is not known. But the erection of its two circles of sarsen stones, apparently during the building phase of Stonehenge II when that monument's two circles of bluestones were put in place, may be significant.

Best-known of Stonehenge's neighbors, and an archaeological site of prime interest itself, is the huge complex of stones at Avebury. This great structure, some 17 miles north of Stonehenge, has suffered far more than its southern counterpart because of its location. It lies under and around houses, streets and fields of the village of Avebury. Almost all of its stones are missing—many are parts of the walls of the quaint thatched cottages of the village—but enough remain to give some idea of original patterns. Most of the enclosing bank and ditch are discernible, particularly from the air.

The Avebury monument seems to have been "discovered," after having been lost to recognition as an ancient structure, by the sharp-eyed John Aubrey. ". . . I never saw the Country about Marleborough, till Christmas 1648," he wrote; at that time, "the morrow after Twelf day, Mr. Charles Seymour and Sir William Button of Tokenham (a most parkely ground, and a Romancy-place) Baronet, mett with their packs of Hounds at the Greyweathers. These Downes looke as if they were Sown with great Stones, very thicke; and in a dusky evening they looke like a flock of Sheep: from whence it takes its name. One might fancy it to have been the Scene where the Giants fought with stones against the Gods. 'Twas here that our Game began: and the chase led us . . . through the Village of Aubury, into the Closes there: where I was wonderfully surprised at the sight of those vast stones, of which I had never heard before; as also at the mighty Banke and Graffe [ditch] about it. I observed in the Inclosures some segments of rude circles, made with these stones, whence I concluded, they had been in old time complete. . . ."

Aubrey thought it "very strange" that "so eminent an Antiquitie should lye so long unregarded by our Chorographers," and soon after the Royal Society was founded in 1662—Aubrey was one of the original Fellows—he wrote that three other members, King Charles II, Lord Brouncker and Dr. Charleton, were "discoursing one morning . . . concerning Stoneheng," and "they told his Majestie, what they had heard me say concerning Aubury, sc. that it did as much excell of Stoneheng as a Cathedral does a Parish Church. His Majestie admired that none of our Chorographers had taken notice of it: and commanded Dr. Charleton to bring me to him the next morning." Aubrey showed the king a "draught of it donne by memorie only" which so interested Charles that when next he went to Bath he "diverted to Aubury, where I shewed him that stupendious Antiquity. . . ." As his Majestie "departed from Aubury to overtake the Queen he cast his eie on Silbury-hill, about a mile off: which he had the curiosity to see, and walkt up to the top of it. . . ."*

Aubrey made surveys of Stonehenge and Avebury and "composed" the "discourse" on them which was quoted from in Chapter 1. He thought that both of them, along with other such monuments, were druid temples, and he never lost his interest in them, but when the king "commanded me to digge at the bottom of the stones . . . to try if I could find any human bones . . . I did not doe it." (Would that some later antiquaries, requested to excavate ancient sites, had behaved similarly!)

The Avebury monument is so disguised by the town and has suffered so from rock-robbers and vandals that if Aubrey had not detected it, the huge, sprawling structure might have been lost forever. In a fashion, Avebury has revenged itself on its destroyers, however. Recent rebuilding of its great circle turned up bones of a man who had apparently been crushed by a falling stone as he was helping to topple it over. Coins in his purse indicated that he lived, and died, in the fourteenth century. And the scissors and lancet found with him showed that he was probably that happy medieval combination, a barber-surgeon.

Avebury apparently began as two stone circles, each about 320 feet in diameter, their outer edges some 50 feet apart, their centers on a north-northwest–south-southeast line. There may have been parts of a third circle of the same size 100 feet to the north and on the same axis line. In this first stage of construction there was probably built most of the avenue also. This concourse is 20 feet wide, bordered by sarsen stones, and runs from Avebury to the Sanctuary. Of the two circles, the more northern, called the "central circle," contained 30 standing stones. Only four survive. Near its center there seems to have been an odd structure called "the cove"—three huge stones set as three sides of a square with the open side to the northeast (though not on the midsummer sunrise line). Only two of these stones are still standing. There are similar "coves" in Somerset and Derbyshire. Their purpose is not known. The "south circle" of Avebury seems to have been slightly larger than the central one. It was made of 32 stones, five of which survive, with a 21-foot-long upright in the center

---

* Silbury Hill will be described later in this chapter. The most remarkable thing about it in the eyes of King Charles was its snail population . . . happening to see "some of these small Snailes . . . no bigger than small Pinnes-heads, on the Turfe of the Hill," he was so surprised that he ordered Aubrey to pick some up, and "the next Morning as he was abed with his Dutches at Bath, He told her of it: and sent Dr. Charleton to me for them to shew Her as a Rarity."

and possibly other stones near the center. The third or "north circle" —if it ever existed at all—may have been incomplete. Excavations carried out in 1960 seem to have established definitely that a third complete circle did not exist.

The second wave of building at Avebury brought demolition of whatever there was of that north circle. The builders set up a north-south line of smallish stones—"smallish," that is, by megalithic standards—in the south circle. They erected a single stone, called the "ring stone" because of a natural hole, outside of that circle to the south. Around both central and south circles and cutting through the site of the north circle these phase II builders at Avebury dug an enormous steep-sided flat-bottomed ditch, roughly circular, with a diameter of some 1250 feet. Outside this ditch and separated from it by a 15-foot ledge the chalk rubble was piled into a large bank. Just inside the ditch was erected a "great circle" of about 100 giant sarsen stones, the largest weighing over 40 tons. This vast ditch-bank-stone ring, some three times as wide as the Stonehenge ditch-bank circle, was quartered by four entrances. The old avenue from the Sanctuary was joined to the southeast entrance by an abrupt, awkward bend.

Both Avebury phases were probably contemporary with the Stonehenge building. Broken bits of pottery found in excavations at the site and two Beaker graves found at the bases of stones of the Avebury-Sanctuary avenue indicate that this tremendous structure was erected about 1750 B.C.

The stones of Avebury are remarkable in two ways. They seem to have been shaped naturally with no tooled dressing, such as distinguished the later Stonehenge stones, and they seem to have been placed alternately in two basic shapes—tall with vertical sides, and broad and diamond-shaped. It is thought that perhaps these two shapes symbolized the male and female principles and that their careful selection and alternation show that the builders were honoring some fertility cult. (Many of the undressed stones of the Stonehenge II double bluestone circle were also of similar shapes, as were the two bluestones, 31 and 49, which now flank the entrance. But the shaping seems to have been fortuitous and the relative placement irregular.) It is also thought that Avebury was the most important temple-meeting-place in the area and probably in the whole of the British Isles—until Stonehenge supplanted it, perhaps in part literally as well as symbolically. There seems to be a strong probability that some of the stones which were first erected at Avebury were later taken down, hauled over to Stonehenge and re-erected there. Such a dismantling of an older monument to furnish material for a successor was not uncommon in Britain, and it would certainly seem reasonable in this situation of two similar structures both requiring huge stones, located only 17 miles apart.

In any case, Avebury was obviously a site of the utmost significance, which apparently yielded precedence to its southern neighbor, and, if so, would have passed on to Stonehenge building concepts and experience as well as actual stones. Avebury has not been excavated as thoroughly as has Stonehenge; further investigation of the 28 acres of this enormous monument may shed valuable light on many of the design problems which presently puzzle the Stonehenge analysts.

There are two main reasons why this larger, perhaps older, and in some ways equally interesting monument has escaped the enthusiastic digging which has disturbed the sleep of Stonehenge, and also the above-ground speculation which has revived so much of that monument's past. Avebury's relics lie around and in a town, which makes archaeological excavation difficult, and the surviving Avebury stones, being untooled, have never given the strong impression of man-made mystery which has so increased curiosity about Stonehenge. Indeed, until fairly recently Avebury has aroused very little serious consideration. Whatever secrets of alignment or numbers it possessed are still unknown.

Large as Avebury was, it seems there was a still larger "henge"-type monument close by. About 100 yards north of Woodhenge there are remains which indicate that this site, now called "Durrington Walls," was once a tremendous circle, with a diameter of perhaps 500 yards as compared to Avebury's 420. All that is known is that it included a bank outside of a ditch. No post or stone holes have been detected. Durrington Walls and also Woodhenge lie very close to the main axis line of Stonehenge, a geometric relationship which future findings may prove to be important.

Last of the presently-known major prehistoric structures near Stonehenge is Silbury Hill. A half-mile north of the West Kennet long barrow and sixteen miles north of Stonehenge, Silbury Hill is the largest artificial mound in Europe. One might call it the great pyramid of Europe. It is a gently sloping conical mount rising to a height of 130 feet, its base a circle more than 200 yards in diameter. It covers some 5½ acres. It is made of chalk rubble dug from a ditch which lies around its perimeter to the north and extends out a long way westward. This ditch was originally about 20 feet deep. Only the top three-fourths of the mound is artificial—the bottom fourth is the north end of a natural ridge of chalk which was used as a foundation. This ridge was cut away on the south to make the mound's shape conical.

Nearly half a million cubic yards of chalk had to be dug and hand-carried to create this vast mound, which means that the effort required about two million man-days of work—a total perhaps greater than that required for the building of the whole of Stonehenge.

Although far larger than any other known barrow, circular or long, Silbury Hill resembles a giant barrow more than it does anything else, and it has been romantically supposed that the hillock might have marked the tomb of some superlatively powerful Stone Age king. But as yet there has been no evidence to support this intriguing theory. In 1777 a shaft was dug from the top straight through the mound to the underlying chalk and nothing was found. In 1849 a tunnel was dug in from the south side to the bottom of that vertical shaft, and again nothing significant was found.

At present, the purpose of this monstrous earthwork remains a complete mystery. So does its date. It *may* have been contemporary with Stonehenge. Indeed, the theorists who suppose that it *may* be the greatest of British tombs sometimes go a little further and suppose that the great man it memorialized *may have been* the architect-designer of the greatest of British prehistoric monuments—Stonehenge.

That supposition may not be irresponsibly fanciful. As this chapter has shown, the area around Stonehenge was obviously regarded by Stone and early Bronze Age men as of supreme importance. There they conducted ritual services and gathered for other purposes which we can only guess at now; there they worshiped, and buried their dead. Is it not conceivable that the energetic, efficient builders capable of erecting such huge structures as Avebury and Stonehenge almost simultaneously would raise a sepulcher worthy of the man responsible for the planning and carrying-through of their culminating creation?

We know what ingenious precautions the pyramid builders took to hide the tomb chambers from the anticipated would-be grave robbers of the future. Could the Silbury Hill creators have tried for such concealment? Might excavations some day bring to light there the tomb of some Stone Age Daedalus?

A thousand years after the megalithic builders of Britain had laid down their tools, leaving monuments and memories long centuries older than Homer's Troy, the Greek poet Pindar wrote, "Neither by ships nor by land canst thou find the wondrous road to the trysting-place of the Hyperboreans." The Hyperboreans were semimythical people who lived far to the north of Greece—we will meet them again in Chapter 8. Pindar's word for "trysting place" was ἀγῶνα, which can mean a gathering place for sports, trials, battles or other activities. Was his Hyperborean ἀγῶνα a racecourse or parade ground like the Cursus, an enclosure like Woodhenge or the Sanctuary, a great open circle like Avebury, an eminence like Silbury Hill, a cathedral-court-observatory like Stonehenge—or all of them? What road could be more wondrous than that which led to the complex, magical trysting place of the great monuments of Salisbury Plain?

Of course, Salisbury Plain was not the only important location of prehistoric tombs and megalithic monuments in Europe. All the way from northern Scotland and Ireland to the Mediterranean there were such structures. Most of them displayed marked similarities of design and construction and many of them were nearly identical. It cannot be overemphasized how much flow and interchange of people and ideas there was throughout the whole of the known Western world in those ages. Moving about must have been unimaginably difficult and dangerous, particularly when there was open water to be crossed. Two thousand years later the sea was still such a menace that a poor anonymous seafarer, battered by the "fearful roll of the waves" in some tempest-tossed vessel and "numb with care," described sea-voyaging as "the road of the wretched." And as late as the seventh century A.D. the Archbishop of Canterbury had to wait in Paris for the whole of one winter before he could cross over to England. Nevertheless, our neolithic ancestors, perhaps assisted by a narrower North Sea and a warmer climate, managed a surprising amount of travel. And it was not all for the purposes of battle, trade and/or mass migration. Along the "road of the wretched" and its possibly less miserable dry land extensions there traveled priests, architects, builders.

I have stood in the great circle of Avebury near the southern end of that extraordinary prehistoric road called the Icknield Way and tried to imagine the appearance of the voyagers along that 200-mile artery which ran all the way from Salisbury Plain up to the Norfolk coast above London, widening in places into an ancient equivalent of a modern four-lane superhighway. I have not succeeded. Why would primitive people, possibly without wheeled vehicles, build such a wide turnpike? What sort of traffic moved along it, that broad highway, all those centuries before the Romans laid out their straight and narrow roads, another thousand years before Chaucer's pilgrims jostled each other along those winding country lanes to Canterbury? All that one can know is that on such ways passed such men and such spirit that there arose throughout the land those memorials to death and life which have so long outlived their creators.

England, Wales, Scotland and Ireland are dotted with hundreds of monuments, burials and stone circles. Only a few have been investigated, but it seems that none are as elaborate as those of Salisbury Plain.† The nearest megalithic structures comparable in subtlety and interest are some 240 miles to the southwest, over the sea, in France.

On the south coast of Brittany, at the base of the Quiberon peninsula near Locmariaquer whence come the best oysters in the world, is the little town of Carnac. It is no kin to the famous Karnak of Egypt, site of the temple of Amon-Ra (which faces the midsummer sunset). But it does lie among a forest of strange and ancient stones.

Less than a mile to the northwest of Carnac, at Menec, is a huge semicircle of 70 closely-spaced stones. Leading to the semicircle from the southwest is a column, 100 yards wide and 1100 yards long, formed of 11 parallel rows of almost 1100 menhirs. (Menhir, from the roots "men" for stone and "hir" for long, means a single large stone. Dolmens or cromlechs are stone structures formed by uprights capped by lintels. Only at Stonehenge is the unit of two uprights capped by a lintel called a "trilithon.") These menhirs increase in height from 2 feet to 12 feet as they approach the semicircle. The visual effect is shattering. It is as if one were looking at an army, fatal, invincible, eternal, marching—and growing as it marched. No wonder local legend says those stones are petrified Roman soldiers. Old wives' tales embroider the legend to the effect that on Christmas night the spell is lifted and the green-gray figures of granite move down to the river to drink.

Some 350 yards to the east-northeast of the marching men of Menec is Kermario, "the place of the dead." There is another petrified army: ten rows of a thousand stones, forming a column 1300 yards long. These rows point to a dolmen and a barrow at nearby Kercado. Another 100 yards to the east-northeast is Kerlescan, "the

† A megalithic monument at Callanish in Scotland has recently been found to be of considerable interest, but the results of investigations of that site were discovered too late for inclusion here. They appear in an article in the appendix.

place of burning." There the army has 13 rows in a column about 900 yards long and 140 yards wide. But only 555 menhirs remain to mark it. Like those at Menec, these rows lead to an irregularly-shaped circle which encloses a gallery grave covered by a mound bordered by stone slabs. One tall menhir stands above the grave.

All three of the columns are oriented northeast-southwest.

It is thought that the Menec-Kermario-Kerlescan structures may have formed a single colossal system. It seems probable that they were built by the same groups of people, some amalgam of culturally similar folk-nations in touch with Britain and other lands to the north, Spain and the Mediterranean to the south and east. Excavation has shown that in addition to being busy travelers and traders these builders slaughtered horses and cattle in their funeral rituals, but little else can be deduced concerning them. At present, even their dates are not well established. Estimates of the probable time of construction of these stone armies of the Morbihan region vary from considerably B.C. to a little A.D.

Further investigation may discover much of intense interest at Carnac, and other megalithic sites in France, Spain, Corsica, Malta, Italy, Crete, and Greece . . . and at Stonehenge, too, for that matter. As this book is being written in late 1964 there comes news from England: within three-quarters of a mile from the center of Stonehenge a Scottish archaeologist, Miss E. V. W. Field, has found a deep shaft. First accounts describe it as a 20-foot funnel-shaped depression tapering into a hole 6 feet wide and "at least 100 feet" deep. The hole contained bits of Bronze Age earthenware. Markings on the walls suggested that the digging instruments may have been bronze tools or antler picks.

A shaft 6 feet wide, dug 100 feet down into the solid chalk . . . what in the world, or under it, could that have been?

Chapter 6

FIRST THOUGHTS

As a boy in England I took little enough interest in my country's most famous ancient monument. I knew that it somehow pointed to midsummer sunrise, and I thought that the druids had built it, probably for human sacrifice, and beyond that my curiosity did not go. Actually, I grew up in Great Yarmouth, home of David Copperfield's Peggotty, and was much more curious about the mechanics of how the Peggotty family lived in that upturned boat.

Then I became an astronomer, and began to wonder about the midsummer sunrise alignment.

In 1953 I worked at the Larkhill Missile-testing Base just a mile north of Stonehenge. The idea of a missile-firing base so close to the stones naturally worried many people, but the missiles were always fired safely to the north. There is a story that during World War I a British airstrip commander had complained that the megaliths constituted a hazard to his planes, and formally requested that they be flattened, but I think that story is apocryphal.

From Larkhill I went often to Stonehenge, and soon became so interested that I took to reading about it. I quickly found that there is an immense amount of literature on the subject—so much that I would not presume to add to it now if I did not have new light to throw on the old mystery. Mythologists and sociologists and historians and other specialists as well as archaeologists—and poets—have written about the unique place, in many different ways. However, my attention quickly focused on that one astronomical aspect, the fact, first noted by W. Stukeley in 1740, that the main axis of the monument was aligned to the midsummer sunrise. That seemed to me by far the most remarkable thing about the whole structure.

I was not alone, of course, in my interest in that alignment. The sad fact is that the fame, or notoriety, of viewing midsummer sunrise over the heel stone has grown to such proportions that thousands of people come each year to watch, and to carouse. Each June an increasingly carnival-like air pervades the site, beginning the night before the sunrise itself. So many merrymakers gather that occasionally the great event is marred by near-riot. The June 22, 1956, *Salisbury and Winchester Journal* reported thus: "Fifteen military policemen were called out . . . yesterday . . . to restore order at Stonehenge where fireworks and an unruly mob threatened to prevent the Druids from carrying out their Annual Summer Solstice Ceremonies. . . ."

The sunrise alignment has interested other astronomers. Since the line from the center over the heel stone does not *exactly* point to midsummer sunrise today, earlier astronomers assumed that the error had been caused by time—that is, by the slow drift of the horizon point of midsummer sunrise during the centuries since Stonehenge was built. Because the angle, or "tilt," of the earth's axis with respect to its orbit plane changes with time, the point on the horizon at which the sun rises on midsummer morning moves, very slowly. For the last 9000 years this movement has been to the right along the horizon at a rate of about $\frac{2}{100}$ of a degree per century. Since this motion can be calculated very accurately, and since it seemed reasonable to suppose that the Stonehenge builders had aligned the monument to point exactly to midsummer sunrise, it was thought that the date of building might be deduced by determining when the axis had pointed to midsummer sunrise.

In 1901 the brilliant British astronomer Sir Norman Lockyer* made such a determination, and arrived at an estimated Stonehenge construction date of between 1880 and 1480 B.C. As we have seen, that estimated date was quite close to the actual date (circa 1850) —but Lockyer's result was discredited when it was announced, because two of his basic assumptions were not accepted as unique or even compellingly probable by archaeologists:

1) He assumed that "sunrise" was the first flash as the top of the

---

* Lockyer (1836–1920) was an extraordinary man whose true worth as an astronomer and theorizer concerning the history of astronomy has not yet been adequately appraised. As a result of observations of the solar eclipse of 1868 he and chemist Edward Frankland independently discovered a new element in the sun's chromosphere which was named "helium," from the Greek word for sun—27 years before that element was discovered on earth. Crediting Henrik Nissen of Germany with the first suggestion (made in 1885) that ancient structures might have astronomic orientations, Lockyer after 1890 attempted to establish such orientations for the pyramids and other antique monuments. Not all of his work has been proved valid, and presently he is in disfavor, but his ideas concerning astronomical orientations remain seminal and I for one agree with the pronouncement made by M.I.T.'s Giorgio de Santillana in his preface to the 1964 reprint of Lockyer's *Dawn of Astronomy:* "The time has come . . . to honor Lockyer as a pioneer, and to carry on in his spirit, with securer data."

sun appears over the horizon, but, the archaeologists pointed out, modern man does not know whether ancient man regarded sunrise as first flash; or midpoint, when the disc's center appears; or "last flash," as the whole sun lifts clear of the horizon. The differences between the three positions are large—at Stonehenge, on midsummer day, the angular distance between the horizon points of first flash and final disc clearance, four minutes later, is almost a full degree.

2) Lockyer assumed that the Stonehenge builders had aligned the line from the center to the Avenue midpoint to point to the sunrise; if he had made the equally plausible assumption that they had intended the center-heel stone line to point to the first flash of the solstice sunrise, he would have produced an estimated construction date of about 6000 A.D.!

In this connection Petrie made an odd mistake in his 1880 book about Stonehenge. He wrote, "There can . . . be no doubt that the first appearance, and not the middle or completion of sunrise, was to be observed, as only the first appearance could coincide with the Heel stone at any possible epoch of erection," basing this conclusion on his assumption that the "obliquity of the ecliptic is decreasing . . . the sun at the solstice has risen . . . more easterly than now . . . the sun's azimuth of rising is decreasing. . . ." Actually, of course, the obliquity of the ecliptic is decreasing, but the effect is the opposite of what Petrie thought—the sun's azimuth of rising is increasing, which means that its solstice horizon point of rising is moving eastward. He calculated that "the sun rose over the peak of the Heel stone at 730 A.D.," plus or minus 200 years, whereas in fact the first flash will not occur over the heel stone for several thousand years.

Since Lockyer's time there had been little direct astronomical investigation of Stonehenge, although the problem of the solstice sunrise alignment continued to be of concern to those astronomers who interested themselves in the monument.

In 1960, I was writing a book on astronomy, *Splendor in the Sky*. In a discussion of eclipses, and the ancients' attitudes toward them (terror, mostly—even after the cause was understood), I wrote, "There must be a great deal of magic that has been forgotten in the course of time . . . Stonehenge probably was built to mark midsummer, for if the axis of the temple had been chosen at random the probability of selecting this point by accident would be less than one in five hundred. Now if the builders of Stonehenge had wished simply to mark the sunrise they needed no more than two stones. Yet hundreds of tons of volcanic rock were carved and placed in position. . . . Stonehenge is therefore much more than a whim of a few people. It must have been the focal point for ancient Britons. . . . The stone blocks are mute, but perhaps some day, by a chance discovery, we will learn their secrets."

As I wrote those words I suddenly thought, "some day" perhaps is now—what better time for that "chance discovery"? I felt that the astronomic aspect of Stonehenge should be thoroughly explored.

By then I had gone from England to Cambridge, Massachusetts, to continue research and teaching. My wife and I made our plans, and the following summer we returned to England, like hunters stalking Stonehenge's celestial secret.

Like proper hunters, or explorers, we set up our base camp in a hotel in Amesbury, close by, and checked our equipment: cameras, compass, watch, binoculars, astronomical tables. Many people came that year to see the sunrise, but few could have prepared for it so meticulously. We had deliberately planned our visit for June 12, nine days before the solstice, because we feared that on the day itself the crowd would make it impossible to set up a camera on the correct alignment and have an unobstructed view, and from previous calculations I knew that the sun would then rise just one diameter to the east of its solstice position.

Dawn was to be about 4:30, daylight time. Among all our welter of preparations the night before we forgot two things: to pay our hotel bill, and to tell the manager that we would be going out so abnormally early. So feeling and looking like the archcriminals the authorities certainly would have branded us had they seen us—one really has to fall foul of it to appreciate the depths and heights of outraged dignity to which English officialdom can reach—we furtively tiptoed down the long dark hall, no sound disturbing the silence except the soft ticking of the grandfather clock. We tried to perform the mechanical feat of starting our car quietly, and we envied the mythical nymph who moved so lightly over the fields that her footfall hardly bent the tassels of waving grain as we glided with a loud crunching sound over the driveway gravel.

Stonehenge stood black and massive against the lightening sky. From a distance it was most imposing. As we looked across the downs we saw not much evidence of dilapidation, and except for the modern road the time could have been June, 1600 B.C. A few hares were scampering around, starlings were chirping loudly, and it was quite cool.

At the site we found that we were not the only visitors. A family from California had spent a cold and miserable night in their Volkswagen bus and were understandably eager to welcome the dawn, and a man passing by on a motor scooter en route from Kent to the north of England stopped, his teeth chattering, to wait for the moment. He was content to see the sunrise as he stood by the road; apparently many people pause at dawn in the general area around Stonehenge.

I set up my eight-millimeter movie camera with telephoto lens trained down the axis line so as to include in its field the sarsen circle archway through which the distant heel stone showed darker than the dark ground. We waited. Purple-tinged mist drifted across the valley, and we were apprehensive lest it creep up Larkhill and obscure the sun. Then suddenly, in the band of brightness to the northeast, we saw it—the first red flash of the sun, rising just over the tip of the heel stone!

It was a tremendous experience. The camera's whirring was the only reminder that we were not in the Stone Age; we experienced primitive emotions of awe and wonder.

Then, as I returned to the twentieth century and began to walk around, my astronomical sense reasserted itself. I felt strongly that the sunrise line had certainly been carefully planned, and that many other stones had also probably been laid out with alignment intended. Indeed, as I peered over and between the stones, I came to feel that *all* of them might have been placed according to some master plan; their relative positions seemed so carefully arranged. It was as if the stones were posing questions which called out for answers, like these:

1. On midsummer morning the full disc of the sun would rise over the heel stone so precisely that if I had been a Stone Age man I would have been delighted or frightened or comforted or awestruck or whatever the priest-astronomers wanted me to be—that alignment had been beautifully established.

Why?

2. The trilithon archways are astonishingly narrow. The space between the gigantic pillars is so small that you can hardly poke your head through (I tried). The average width of the three standing archways is 12 inches, and the average thickness of the bordering uprights is 2 feet, so that when you look through two aligned archways your view is restricted to a very small angle. I felt that my field of observation was being tightly controlled, as by sighting instruments, so that I couldn't avoid seeing something.

What was I supposed to see?

3. The sighting-lines through the trilithon archways extend on through corresponding wider archways of the surrounding sarsen circle. But as I walked along the axis I noticed that those three sighting-lines flashed into view one after the other, and, as rapidly, out of view again. At no one spot could I stand and look down all of those double-archway-framed vistas. Viewing had to be from well-separated points. Such as arrangement is unusual. It violates customary architectural design which radiates vistas from a central single focus, and it somehow seems not "natural." I felt again that the placement had been deliberate, to stress the importance of the viewing.

Why was the viewing important?

4. The only two outer stones now standing, number 93 and the heel stone, are both of such a height that an average-sized man looks across their tops to the line of the horizon.

Why was there such precise arrangement of height?

5. The line joining corners 91–94 of the station stone rectangle lies just a few feet outside the stones of the sarsen circle.

Did they form a sighting-line which had been preserved?

Most of those questions, I felt, might somehow be answered by astronomy. Those precise alignments and controlled vistas, so carefully directing the eye to nothing now visible, might well have been sighting-lines for celestial events such as special rise or set points of those godlike forces of prehistory, the sun, moon, planets, and stars. Primitive men observed with apprehension the places where the great rulers of day and night entered and emerged from the dark earth. It would have been natural that the Stonehengers should mark those points by various means.

I thought immediately of the most obvious "God," the sun. As most schoolboys and all sailors, farmers, navigators and astronomers know, the sun moves from north to south as June moves to December. Only two days in the year—the spring and fall equinoxes—does it rise and set due east and west. Because of heavenly complexities involving factors like the obliquity of the ecliptic, which it is fortunately not necessary to discuss here, the sun swings annually from a summer declination (or celestial sphere latitude) of $+23.5°$ (north) to a corresponding winter declination of $-23.5°$ (south). That declination shift is a sizable $47°$, but because of the facts of spherical geometry the angular variation in earthly viewing can be much larger. At the latitude of Stonehenge sunrise goes from a compass direction

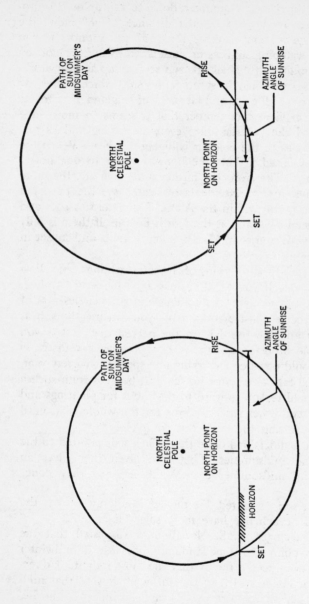

Fig. 9. The daily journey of the sun for a latitude of approximately 40° north. The spin of the earth on its axis causes the sun to appear to rise, move in a circle around the north celestial pole, and set.

Fig. 10. The daily journey of the sun for a latitude of approximately 60° north. At high latitudes the circular path of the sun appears to be higher in the sky and consequently the rise and set positions are closer to the due north point on the horizon.

of 51°, almost northeast, at midsummer, down to 129°, almost south-east, at midwinter. That is an angular distance of 78° along the horizon, an average motion of more than 12° per month. If you have the habit of watching sunrises or sunsets, you will have noticed the astonishing rapidity with which the sun seesaws up and down the sky. And if it seems odd that in summer the sun, which everybody knows is always south of Florida and far south of England, rises to the north of an English viewer, remember that it seems to move in a small circle around the polestar once every 24 hours, and as one moves north on the earth the polestar is higher overhead. When the path of the sun is raised, it cuts the horizon closer to due north. (See Figs. 9 and 10.) Therefore, the farther north you are, the more northerly is summer sunrise. Residents of Alaska see the June sun rise practically due north; within the Arctic Circle the sun rises and doesn't set for several days, and at the North Pole itself there is only one "day" a year, with sunrise in March, noon in June, and sunset in September.

By means of this north-south swing of the sun earthlings can follow the course of the year. If you are a sophisticated modern earthling, with knowledge of latitudes and declinations and great circles—and if you have some rather expensive equipment—you can use the sun as a cosmic calendar and tell the date to the nearest day. But if you were only a simple Stone Age man, you might regard yourself as fortunate if you could be sure of marking one special day every year, and you might well take great pains to mark it, because from such a known day you could reckon forward to the times for plantings and harvests, hunting, and other vital concerns for the whole year, until that day came again and the cycle was complete.

The Stonehenge builders had done that. Their axis pointed to the place of sunrise at midsummer. They had given themselves an accurate marker for midsummer day. What else had they done?

I thought of the sun, as its red disc moved rapidly away from the heel stone. Could Stonehenge have more solar alignments?

The noted archaeologist R. S. Newall once suggested that the axis reversed might point over some landmark, now lost, to midwinter sunset. There has even been a theory that the most important direction of Stonehenge was intended to be southwest, toward that midwinter sunset, rather than northeast, toward midsummer sunrise, because the Avenue entrance is from the northeast and most structures, like cathedrals, have the most important direction opposite the entrance. But that theory has not been proved. Nor has evidence ever been found that there was a marker on the axis extended toward the southwest.

Could there be alignments to celestial bodies other than the sun—to the stars or planets or moon?

The sun was moving eastward at such an angle that it was a full degree to the right of its first flash position when it finally lifted clear of the horizon. I marveled once more at the precision of placement of the axis and the heel stone, and at the whole precision of Stonehenge. I kept looking at those alignments formed by the ancient stones, and thinking of the many objects in the sky, and suddenly I felt defeated.

"It's no use just wondering," I said to myself. "To answer these questions—to find if these alignments have any celestial significance—we need precise measurement and comparison, a great volume of trial-and-error work—much more work than I can find time to do.

"We need the machine."

Chapter 7

THE MACHINE

Computers are indeed wonderful things.

They are, of course, not new. For about as many ages as he has been Homo sapiens, perhaps for *exactly* as many ages, man has used things as tools to help him count. First there were fingers. Then, sticks, stones, scratches, any units which could be grouped and tallied. Then more elaborate devices like the sandglass, the running-water clock, the 2500-year-old abacus (which, in the hands of a good operator, is still faster than an electric desk calculator). The ancient Chinese also used small "counting rods," and the Romans made simple computations with little pebbles, or "calculi." The tenth-century Pope Sylvester II was credited with magical powers of divination, possibly because he mastered the abacus which the Saracens were then using. Three hundred years later the learned Roger Bacon developed many ingenious engines, some of them perhaps capable of performing calculations—he was popularly supposed to have obtained prophecies by means of a brazen head. In the sixteenth century Lord Napier, inventor of logarithms, apparently performed arithmetical and geometrical calculations with "certain pieces of wood or ivory with numbers on them, and these were called Napier's Bones." And in the seventeenth century the art of mechanical computing began to become a science.

In that century England's William Oughtred invented the slide rule. (Oughtred was the gentle cleric who taught Christopher Wren mathematics. Aubrey said he was a "pittiful Preacher" because he "bent all his thoughts on the Mathematiques . . . his head was always working. He would drawe lines and diagrams on the dust," but he could "bind up a Bundle well" and as an astrologer he was "very lucky . . . his son Ben was confident he understood Magique.") France's Blaise Pascal designed a set of wheels "for the execution of all sorts of arithmetical processes in a manner no less novel than convenient." And Germany's Leibnitz made a crude device that could multiply.

At the end of the next century the French tried to make a monstrous calculating machine out of about a hundred human beings, but even Napoleon couldn't order that. In the nineteenth century the extraordinary Englishman Charles Babbage, responsible for dozens of innovations including flat-rate postage, skeleton keys and the cowcatcher, put together a "Difference Engine" which managed to compute simple mathematics tables. Then he dreamed, publicly, of an improved "Analytical Engine," capable of performing at the then alarming rate of sixty arithmetic operations a minute. The idea of that machine attracted many supporters—Byron's daughter Ada Augustus, Countess of Lovelace, was an ardent backer (she was a surprisingly good mathematician). But the "Analytical Engine" never got off the drawing board. After Babbage, there was little improvement in the machine calculation field; Victorian computers were turned by hand, at a suitably stately pace.

The really great advance took place in the 1940s. Howard Aiken of Harvard, employing some of the principles of the old "Analytical Engine," devised an automatic sequence controlled electromechanical computer. His "Mark 1" was completed in 1944. The next year John von Neumann proposed internal storage, and the race was fairly on. Now, a scant twenty years later, those early collections of vacuum tubes, switches and flashing neon bulbs have metamorphosed into transistorized magnetic tape giants, which shape the world of our time, and beyond.

A modern electronic digital computer like the IBM 7090 has 50,000 transistors, 125,000 resistors and 500,000 connectors, joined by some twenty miles of wire. Its successor, the 7094, has about 10 per cent more of those components, and is about a third faster in operation. The next generation of machines will be faster still. (And, oddly enough, the machines are growing smaller—because of increased use of transistors and other miniature parts, and more efficient circuitry.) A typical computer consists physically of about twenty units—tall cabinets filled with calculating and recording devices, many with two tape reels visible, behind glass, at the top. It requires about 45,000 volt-amperes of electric current, about 70 horsepower.

It can perform 250,000 simple operations—additions, subtractions, trigonometric functions, etc.—per second, producing its answers in lines containing 26 5-unit "words" in figures or in alphabet letters or in any other code you choose, at the rate of 600 printed lines per minute. At those rates it could "read" the whole Bible in a minute, print it in some seven hours. It is uncomplaining, untemperamental, tireless—like that of mercy, its quality is not strained, nor is its capacity. Furthermore, it does not make mistakes.

In the early days of the model called "650" we were told that certain slight errors in a numerical check were caused by the machine's "warming up." We believed that. And we were wrong. The machine was trying to tell us that there was a significant error in the program that we had put into it; ultimately we had to re-calculate the entire program. Nowadays if there is an error in the input program the computer not only detects it but gives the approximate description and location of the error and recommends procedure for correction. I am told that for new programmers this can be rather unnerving.

Computers are now being used for a wide range of tasks including such not obviously mathematical jobs as weather forecasting, diagnosis of illness, invention, literary composition and translation. In our space effort they are of course indispensable; without them there could hardly be a space effort. For example, consider October, 1957, when the Russians launched the first artificial satellite. At that time the best computer, the "650," worked at the now-primitive but still not-sluggish rate of 4000 operations per second—but even so there were so many factors involved in calculation of the satellite's motion that the machine took 30 minutes to compute its orbit and thus follow it. The satellite itself, moving at a speed close to 18,000 mph, went around the world in about 90 minutes. The machine had only 60 minutes leeway; if that extra time had been consumed in repair or maintenance the satellite would have been moving faster than the machine was following and might theoretically have been lost. Or if there had been other satellites, the machine would have been swamped. Now there are some 500 man-made objects moving through space—all of them being tracked comfortably enough by the improved machines. The so-called Space Age might just as well be termed the Computer Age.

Present computers also play. They play championship checkers or bridge or such-like uncomplicated games, and they are capable of passable, "barnyard" chess. (Ultimately, with better programs, they will play championship chess too, it seems. Then all the fun might go out of that game—but, say the present programmers, the machine might replace the old game of chess with a new version so complex it would tax the new programmers—a sort of cosmic chess played in three dimensions.) They can figure the odds. A recent movie showed a computer breaking the bank at Monte Carlo, and in actuality a young physics professor with a homemade machine was on his way to disrupting the whole fabric of Las Vegas until he was defeated by defeat itself—the panicking gambling houses surrendered and refused to play with him any more.

Ours is becoming a computer world. University students are nudged into the computer room in their freshman year. To them, the machine is a way of life. Recently I asked a student to do a mathematical job worth about three pencil-hours. A week later she gave me the result. She had referred the problem to the 7090, which meant that for days she had to wait her turn for the use of a fraction of a second of the machine's time. In honest puzzlement I asked her, "Why didn't you use a desk calculator?" "I don't know how." "Then what about a pencil and graph paper?" "What's graph paper?" The moral, I suppose, is that one should keep one's problems hard.

Presently it is a popular occupation among the computer fraternity to compare their mechanism to the human brain. The conclusions are not disheartening—marvelous as the machines are, the brain seems still a good deal more marvelous. Like the mills of the gods, it grinds slow compared to the machines, but it grinds exceeding fine —it is original, imaginative, resourceful, free in will and choice. The machine operates at a speed approaching that of light, 186,000 mi. per sec., whereas the brain operates at the speed at which impulses move along nerve fiber, perhaps a million times slower—but the machine operates linearly, that is, it sends an impulse or "thought" along one path, so that if that path proves to be a dead end the "thought" must back up to the last fork in the road and try again, and if the "thought" is derailed the whole process must be begun again; the brain operates in some mysterious multipath fashion whereby a thought apparently splits and moves along several different paths simultaneously so that no matter what happens to any one of its branches there are others groping along. And whereas even a transistorized computer has a fairly modest number of components, the brain, it seems, has literally billions of neurons, or memory-and-operation cells. To rival an average human brain a computer built by present techniques would have to be about as big as an ocean liner, or a skyscraper. And even then it would lack the capacity for originality and free will. To initiate free choice in a machine the operator would have to insert into its program random numbers, which would make the machine "free" but uncoordinated—an idiot.

In the future, improved computers may cooperate with humans to produce more elegant programs which may in turn enable those computers to come closer to real "thinking," and to approximate brain processes in other ways. Already, it seems, analogies between machine and brain are being suggested. For instance it is indicated (by C. R. Evans and E. A. Newman in the *New Scientist*, November 26, 1964) that the process by which a machine disposes of obsolete, redundant or otherwise useless program instruction—an erasing or sidetracking process done during the machine's off-duty hours —may be akin to human dreaming, which has been tentatively established as a process by which the brain during its off-duty hours examines, arranges, files the good and discards the useless information it has received.

It is certain that future computers will be much more than tools. They will be specialized and organized, vertically from general-purpose "slaves," and horizontally across continents and oceans. They will record, dispose, operate, regulate, solve, devise, predict, explain . . . what will they *not* do? It is not difficult to imagine them abolishing money: each person might have a card which he would show whenever exchange was involved—a quick flash to the central "bank" (wherein was stored nothing but figures) would check his credit and update his balance. It is not difficult to imagine them performing certain delicate functions in the body, such as regulation of heartbeat, or brain activity, or metabolism. It is not, in fact, difficult to imagine them becoming so skillfully and beneficially entwined with our brain and body operation that the old specter of the take-over by machine might be exorcised in the actuality of a symbiotic merger, a cooperation between machine and man.

However, enough of this computer contemplation. I am not a computer man. I don't even know exactly how they work. To get answers from a computer, I have to question it through an intermediary, a programmer. All I can say at first hand is that computers *do* work— and I am glad. Because one of them, the Harvard-Smithsonian IBM 704 (now as obsolete as the hand-crank telephone), did something for me I could hardly have done for myself. It found the secret, or *a* secret, of Stonehenge.

In 1961, after I had decided that the problem at Stonehenge was worthy of a computer's attention, I had to fit that problem to the machine: feed it information it could digest, and ask it a question it could understand and answer. The machine requires definiteness.

Many people have wondered about possible astronomic meanings at Stonehenge, but their wonderings have tended to be vague.

In 1740, before he wrote *Choir Gaure,* John Wood theorized that Stonehenge had been a "temple of the Druids sacred to the moon." In 1771 John Smith noticed the solstice sunrise alignment and speculated on number and shape significance. In 1792 a man now identified only as "Warltire" declared that Stonehenge had been "a vast theodolite for observing the motions of the heavenly bodies . . . erected at least seventeen thousand years ago."

In 1793 a Rev. J. Maurice supposed, on mystical grounds, that Stonehenge had been a temple to the sun. In 1829 one Godfrey Higgins stated that the arrangements of the stones represented "astronomical cycles of antiquity," which indicated an erection date of about 4000 B.C. In the 1840s the Rev. Edward Duke noticed that station stone positions 91–92 and 93–94 are parallel to the Stonehenge axis and so align to midsummer sunrise and midwinter sunset. And in 1873 the Rev. Gidley described the method by which the first most important, astronomic alignment had been checked at the site: "Dr. Smith . . . without the aid of any instrument, or assistance, except from a 'White's Ephemeris,' came to the conclusion that at the Summer Solstice the sun would be seen by one standing on the Altar Stone to rise over the Bowing Stone." (An ephemeris, from the Greek word for day, is a table giving positions of heavenly bodies; the heel stone has been called the "bowing stone" because of its tilt.) For his own part, Gidley suggested that it was "not improbable" that four positions, which he failed to identify positively— two of them were probably station stone mounds 92 and 94—aligned to point to midsummer sunset and midwinter sunrise. He also noted that although "some writers" had tried to link the monument to the planets he had found nothing "which directly connects the planets, except perhaps Saturn, with Stonehenge."

Petrie concluded (wrongly) in his 1880 treatise that the station stones 91 and 93 "cannot have any connection with solstitial risings or settings." His comment on solstice activities at Stonehenge almost 100 years ago is interesting: "The large numbers of people that keep up with much energy the custom of seeing the sun rise at midsummer, somewhat suggests that it is an old tradition; and hence that it has some weight, independent of the mere coincidence."

In this century there has been a great deal of conjecture, some of it very acute, about possible astronomic significance at Stonehenge. After Lockyer's 1901 attempt to date the monument by astronomic methods several qualified scholars have speculated about celestial orientations and significances. But their speculations lacked one thing —the calculation. Such theories should be tested mathematically. Figures alone put teeth into any astronomical theory—or, if the theorizer is unfortunate, take the teeth out.

For the machine, I needed something concrete; a well-defined problem, the best data available on Stonehenge, and a clearly stated question. Only with such input could there be effective output, and the question answered.

My question was definite enough: "Do significant Stonehenge alignments point to significant celestial positions?" The requirement of *significance*, on the ground and in the sky, was obvious. There are so many possible Stonehenge alignments—27,060 between 165 positions—that one could be found to point to practically anything in the sky, and, vice versa, there are so many objects in the sky—perhaps literally an infinite number—that hardly any line extended from earth could fail to hit at least one.

To answer that question, the machine needed pertinent information about Stonehenge and the sky.

We proceeded to give it that information.

First the programmers, Shoshana Rosenthal and Julie Cole (Judy Copeland joined us later), took a chart showing the 165 recognized Stonehenge positions—stones, stone holes, other holes, mounds—and placed it in "Oscar," an automatic plotting machine.* Then they placed the cross hairs over each position and singular geometric point like the center and the archway midpoints, pressed the button, and "Oscar" punched each point's X and Y coordinates on a card. The X-Y intersection or origin was arbitrarily set well outside the charted area, in the southwest quadrant, so that all coordinates would be positive.

Then they went to the computer. They primed it with the geographic information—the latitude and longitude of "Oscar's" origin point, the compass orientation of the axes, and the scale—and they instructed it to do three things:

1) extend lines through 120 pairs of the charted points (some pairs, such as neighboring points, were judged valueless as alignment indicators),

---

* Most Machine Age machines are numbered or named for their inventors or mythical persons like Jupiter or Thor—how "Oscar" got its name nobody knows.

2) determine the compass directions or azimuths of those lines,

3) determine the declinations at which those lines going out from Stonehenge would hit the sky. (If the heavenly bodies are regarded as lying in a hollow sphere enclosing the earth then the circles on that sphere corresponding to latitude circles on earth are called declinations.)

I hope this is clear. Perhaps it would help to put it this way: it was as if they told the machine to stand at each of the selected points, look across each of the other points to the horizon, and each time report what spot of the sky—the declination only—it saw.

This priming process, the programming of the machine, took about one day.

Then they gave the "Oscar" cards to a computer operator, who fed them into the machine. In a few seconds it transferred the card information to magnetic tape, scanned the tape, processed the information according to the programmed instructions, and shot forth its result—some 240 Stonehenge alignments translated into celestial declinations. (The 120 pairs yielded twice as many alignments because each line was considered as pointing in both directions.)

That task took the machine less than a minute. It would have kept a human calculator busy for perhaps four months. (To check the machine, Mrs. Rosenthal did one of the computations by hand. It took her four hours.)

And so we had half of the answer to our question. We knew where the important Stonehenge alignments met the sky, the declinations. The next part of the question was, "Were those declinations celestially significant? Did they mark special rise or set points of special heavenly bodies?"

We noticed at once that among the declinations which the machine had produced there was a large number of duplications. Figures approximating + (north) 29°, +24° and +19°, and their southern counterparts, −29°, −24° and −19°, occurred frequently. We decided to see what celestial bodies were close to those declinations.

Quickly we checked the planets. The closest one was Venus, but its maximum declination, ±32°, was not close enough. Why Gidley thought there might have been a connection between Stonehenge and Saturn I do not know; that planet's maxima are now about ±26°, and in 1500 B.C. were about the same.

Then we ran through (nice phrase!) the stars. The six brightest stars are, in order, Sirius, Canopus, α Centauri, Vega, Capella and Arcturus. Of those, only Sirius, the brightest, was near. Sirius is at declination —16°39′ now, but in 1500 B.C. was at about —18°, according to Lockyer—the stars change declination at different rates, their positions as seen from earth being affected by their own actual motion, called "proper motion," as well as the motion of the earth's axis relative to the celestial sphere. Arcturus is now at +19°21′ but in 1500 B.C. was at about +40°—nowhere near the lines of Stonehenge. There seemed no probable significance to the possible star alignments; even if further calculation showed that Sirius worked exactly at some date in the past and one or two more alignments of fainter stars turned up, this is just what one would expect from pure chance. Furthermore even a bright star like Sirius can only be seen at rising under extremely favorable weather conditions. Fainter stars are totally invisible on the horizon. We decided to try the most obvious celestial bodies, those prehistoric deities, the sun and the moon.

This time the result was astonishing. Repeatedly and closely those declinations which the machine had computed seemed to fit extreme positions of the sun—which I had suspected that they might—and also—which I had *not* suspected—the moon. Pair after pair of those significant Stonehenge positions seemed to point to the maximum declinations of the two most significant objects in the sky.

I say "seemed" because at that stage we were using a preliminary search program of no great celestial accuracy. The stone alignments and resulting declinations as produced by the machine were as exact as the original chart allowed, but we did not then have correspondingly precise positions for the sun and moon as of the time of Stonehenge. We were using only rough approximations, gotten by mentally chasing those objects backward 4000 years in time. To verify the apparent correlations we needed precise sun-moon extreme positions as of 1500 B.C.

Back, of course, to the machine.

We gave it the present solar-lunar extreme declinations and the rate of change, and instructed it to determine what the extreme declinations had been in 1500 B.C. At the same time we programmed the machine to calculate the direction of rise and set of the sun and moon. Not knowing what the Stonehengers might have chosen we allowed three definitions: (a) sun just showing, (b) sun's disc cut in half by horizon, and (c) disc standing tangent on the horizon. There is about 1° difference between the direction of (a) and (c), which of course is not very great, but I wanted to determine if possible what the Stonehengers had chosen as their definition.

And now I must try the reader's patience with some more basic astronomy. I must explain a little about the moon.

I have explained that the sun moves from a northernmost maximum position of $+23°5$ declination in summer to a corresponding $-23°5$ extreme southern declination in winter. Just the reverse motion is true of the full moon. It goes north in winter, south in summer. And it has a more complicated relative motion than the sun; it has two northern and two southern maxima. In an 18.61-year cycle it swings so that its far north and south declinations move from 29° to 19° and back to 29°. Thus it has two extremes, 29° and 19°, north and south. This pendulumlike relative motion is caused by the combined effects of tilt and precession of the orbit and it is much too difficult to clarify quickly; even an astonomer has trouble visualizing the processes involved. Here it is only necessary to understand that the moon *does* have two extreme positions for every one of the sun.

To position the sun and moon as of 1500 B.C. took the machine a few more seconds. The declinations it reported were $±23°9$ for the sun and $±29°0$ and $±18°7$ for the moon. The most cursory glance showed us that those declinations were close, very close, to the ones determined by the Stonehenge alignments.

We compared the figures carefully. There was no doubt. Those important and often-duplicated Stonehenge alignments were oriented to the sun and moon. And the orientation was all but complete.

As I have said, I was prepared for *some* Stonehenge-sun correlation. I was not prepared for total sun correlation—and I had not at all suspected that there might be almost total moon correlation as well. For what the machine's figures showed was this:

To a mean accuracy of less than one degree, 12 of the significant Stonehenge alignments pointed to an extreme position of the sun. And to a mean accuracy of about a degree and a half, 12 of the alignments pointed to an extreme of the moon.

As the accompanying diagrams (Figs. 11 and 12) and Table 1 show, not one of the most significant Stonehenge positions failed to line up with another to point to some unique sun or moon position. Often the same Stonehenge position was paired with more than one other to make additional alignments. And of the 12 unique sun-moon rise-set points, only two—the midsummer moonsets at −29° and −19°—were not thus marked.†

---

† The stones which would complete these two alignments should by symmetry be near Aubrey hole 28, but this area beyond the ditch has not been thoroughly excavated.

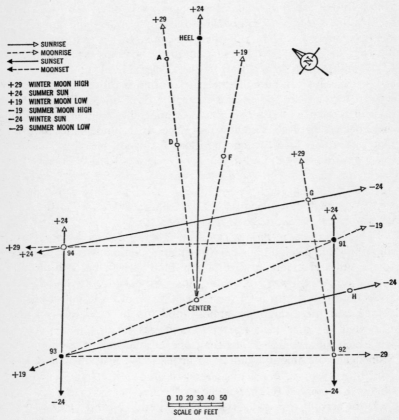

Fig. 11. The original alignments found for Stonehenge I. For precise work the reader should refer to the numerical azimuths listed in Table 1.

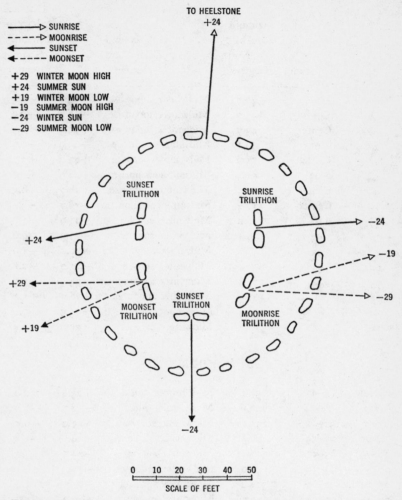

Fig. 12. The alignments found for the archways of Stonehenge III. For precise work the reader should refer to the numerical azimuths listed in Table 1.

TABLE 1

STONEHENGE I

Position	Seen from	Azimuth Clockwise from North Degrees	Object and Declination Degrees		Distance Above or Below Skyline† Degrees
G	92	40.7	Midwinter moonrise	+29.0	−0.5
A	Center	43.7	Midwinter moonrise	+29.0	+0.9
D	Center	43.7	Midwinter moonrise	+29.0	+0.9
91	92	49.1	Midsummer sunrise	+23.9	−0.7
Heel	Center	51.3	Midsummer sunrise	+23.9	+0.1
94	93	51.5	Midsummer sunrise	+23.9	+0.6
F	Center	61.5	Midwinter moonrise	+18.7	+0.3
91	Center	117.4	Midsummer moonrise	−18.7	−3.4
H	93	128.2	Midwinter sunrise	−23.9	−1.3
G	94	129.4	Midwinter sunrise	−23.9	−0.6
92	93	140.7	Midsummer moonrise	−29.0	−1.0
92	91	229.1	Midwinter sunset	−23.9	+0.1
93	94	231.5	Midwinter sunset	−23.9	−1.3
93	Center	297.4	Midwinter moonset	+18.7	+1.2
94	G	309.4	Midsummer sunset	+23.9	+0.3
94	91	319.6	Midwinter moonset	+29.0	−0.4

STONEHENGE III

Position	Seen from	Azimuth Clockwise from North Degrees	Object and Declination Degrees		Distance Above or Below Skyline† Degrees
Heel	30–1	51.2	Midsummer sunrise	+23.9	0.0
8–9	53–54	120.6	Midsummer moonrise	−18.7	−1.2
6–7	51–52	131.6	Midwinter sunrise	−23.9	+0.7
9–10	53–54	139.4	Midsummer moonrise	−29.0	−1.7
16–15*	55–56	231.4	Midwinter sunset	−23.9	−1.2
20*–21	57–58	292.0	Midwinter moonset	+18.7	+5.4
23–24*	59–60	304.7	Midsummer sunset	+23.9	+3.2
21–22	57–58	315.2	Midwinter moonset	+29.0	+1.7

* These stone holes are missing at the present-day Stonehenge and are not marked on any excavation plans. Thus these archway midpoints have been estimated from the symmetry of neighboring positions.
† The "distance above or below skyline" gives the position of the lower edge of the sun or moon, relative to the skyline, at the aligned azimuth. A zero means that the sun or moon stood exactly tangent on the skyline, like a wheel standing on the ground. (See Fig. 13.)

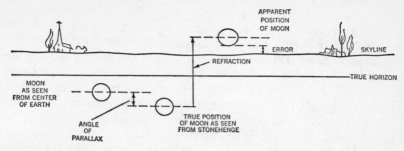

Fig. 13. Conditions at rising or setting. Astronomical calculations are made first for an observer at the center of the earth. To calculate the conditions for an observer at the surface, a correction must be applied for parallax. Then allowance must be made for atmospheric refraction which causes a celestial object to appear higher in the sky than it really is. Finally the skyline altitude must be allowed for because it is usually above the horizontal eye-level surface that defines the astronomical horizon.

The relation between this vertical error and the corresponding horizontal error varies with amplitude of declination. At ±29° a vertical error of 1° means a horizontal error of 1°8, at ±24° the relation is 1 to 1.6, at ±19° it is 1 to 1.5, at ±5° it is 1 to 1.3, and at 0 it is 1 to 1.2.

It will be noted that this table differs slightly from that given in the article "Stonehenge Decoded" which appears in the appendix. That is because after the article was printed, reruns and checks of the machine program refined some of the measurements and added four alignments—the three sun positions 91 from 92, G from 94 and 93 from 94, and the midwinter moonrise G from 92.

It was an extraordinary correspondence.

And the precision of the alignments was noteworthy. The best fit was with the assumption of the sun or moon tangent on the horizon. As the table shows, the average accuracy of the sun lines was 0°8 and the moon lines 1°5. These average errors are caused to a large extent by two "bad" archways with errors of 3°2 and 5°4 on the western side. The error is given in the last column of the table and is shown diagrammatically in the figure. Because of the slanting direction of sunrise, an error of 1° in the vertical direction corresponds to about 1°6 in the horizontal, at 24°.

Usually a scientist does not discuss errors. When all precautions have been taken, an error is recorded without comment because a second attempt might reduce the error and a third attempt cause it to be larger again. An error is an error is an error.

But at Stonehenge we might learn something by such discussion.

Firstly, it will be noted that there are slight discrepancies in the numbers between Table 1 and the table in the appendix. That is because when I wrote the *Nature* article I had no information about actual skyline conditions around Stonehenge and had to assume a uniform skyline—afterwards I obtained a chart showing actual skyline altitude variations around the site. Table 1 therefore gives more accurate figures. However, neither the theoretical uniform skyline nor the actual skyline as of today would necessarily correspond to the skyline that circled Stonehenge in 1500 B.C. Trees growing then where now there are none could have elevated that ancient skyline by some 0°.2—which would mean that an error presently recorded as +0°.2 might actually then have been 0.

Secondly, we found disagreement between one plan and another, and from the data available we were uncertain which plan was more correct. This gives an uncertainty in each figure of about ±0°.2. The error along 94–G might be as small as 0°.1, or as large as 0°.5. This is annoying but not serious. Bear in mind that 0°.5 is a small angle for a naked-eye observer.

Thirdly, some of the trouble may have occurred when the priests were laying down the lines. The sun is easy to see during the several critical days at midsummer and midwinter, and sighting errors would be small. But the full moon had to be observed on *the* night of full moon at *the* particular year of a 19-year cycle. If it was cloudy, and the lines were set the night before or the night after full moon, the moon would not have been exactly at its extreme. When this happened, the error would have been positive when the moon's declination was positive and negative when the declination was negative. A glance at Table 1 will show that this + and − correspondence occurs for 10 out of the 12 moon lines. Perhaps they did have a few cloudy nights!

Fourthly, Stonehenge is not what it used to be. Stones have tumbled over to lie broken or to be re-erected by modern cranes. The worst errors involve stones that have disappeared long ago—24, 15 and 20. For these, I could only make an estimate of the original positions. Perhaps the errors for these three alignments should be left blank until the archaeologists can provide more information. Is there a hole beneath the turf near the expected position, is the hole a foot or two displaced from the estimates that I made? Furthermore, it is just possible that construction was deliberately halted at some stage of the work, because the builders realized that the design problem they had set themselves was insoluble. A completely symmetrical structure could not have exactly fitted the asymmetrical sky positions.

Finally, the most serious displacement of all may be due to modern man. Notice how the moonset archways 57–58, 21–22 are flat in the 1944 aerial photograph. They fell in 1797, before Petrie's accurate survey. The Ministry of Public Buildings and Works pulled them up straight in 1958, but the stones were originally in shallow holes and it was difficult to reset them exactly. My calculations show in the appendix there is a horizontal displacement of 16 inches in one or the other of the archways; perhaps that shift has been caused by the re-positioning of these massive blocks.

Then again the sunset trilithons are presently in a sorry state. The great trilithon is broken, having fallen hundreds of years ago. Although 56 was re-erected in 1901, several authors have questioned the accuracy of the restoration; the stone is not perpendicular to the Stonehenge axis but is turned counterclockwise by several degrees. The summer sunset trilithon is half fallen and the corresponding arch marked by 23 is unreliable. Stone 23 fell, and was finally set in cement in 1964.

To support my suggestion that some of the errors are modern, note that the trilithons and archways which have never fallen are more accurately aligned.

The error for the most famous alignment of all, the midsummer sunrise as seen from the center over the heel stone, deserves particular discussion. At present a six-foot man looking from the center sees the top of the heel stone level with the distant skyline. In 1800 B.C. the first flash of the sun appeared about ¾ of a degree to the north, or left, and so the six-foot man standing in the center would have seen its lower edge pass just one-half of a degree above the top of the heel stone—IF that stone had then been leaning at the angle it stands at today. But if the stone was upright in 1800 B.C., as I believe it was, it stood some 20 inches higher then, and the 0°5 error registered by the machine for its present position would have been practically zero. I have calculated Table 1 on the assumption that the heel stone was upright, and the Stone Age viewer saw the solstice rising sun just graze the tip of the heel stone as it moved upward and over. Here there seems no doubt that the builders intended the disc of the sun to stand exactly on the marker.

Such precision of placement is, or was, astounding. To erect a boulder as irregularly shaped and ponderous as the 35-ton heel stone so that it was horizontally aligned to an accuracy of a foot was a task difficult enough; to sink that great block into the ground just so far and no further, so that its tip was also aligned vertically to an accuracy of inches, was an achievement requiring another whole dimension of skill. How, in fact, was it done? If, after erection, the stone had settled too deeply it would have been out of alignment—and how could it have been lifted? Of course, if it had not settled far enough its top could have been bashed away to lower it to the proper height —but the top was *not* bashed. Perhaps the heel stone was erected first, and the viewing point laid out afterwards?

So much for the errors.

Finally, in a consideration of these sun-moon alignments, it should be remarked how carefully those alignments were preserved, added to, and made more spectacular down through the successive waves of building. During the 300-year period of construction many people of many different thoughts and cultures came to Stonehenge. Different rulers, designers, priests and workmen set their brains and hands to the vast work of alteration, adaptation, change and creation. The great monument grew from a simple circle open toward the midsummer sunrise to a rectangle-within-a-circle to a massive and complex cathedral of stones standing in arched circles and horseshoes. Yet the oldest orientation of all, the axis alignment to summer solstice sunrise, was never lost; rather was it maintained, duplicated, emphasized. Other alignments were similarly maintained and duplicated and made more spectacular. And just as the earlier builders had used every one of the significant stones and positions for repeated alignments on the sun and the moon, the later builders placed their circles and horseshoes so skillfully that not one of the huge trilithon arches failed to align with at least one of the outer circle arches to point to one of the extreme positions of the sun or moon.

What the original builders had done was remarkable enough; to arrange a circle and a rectangle and six outlying stones so that between them, paired, they form 16 alignments on 10 of the 12 unique sun or moon points is very difficult. What the last builders did was even more remarkable; they duplicated 8 of those earlier, two-position alignments in archwayed vistas. Where the Stonehenge I and II people obtained their sighting directions by standing at one place and looking over another, the men of Stonehenge III saw 8 unique sun and moon risings and settings *through* tall stone arches. And the last builders, like the first, used one position for more than one sighting line—see how the two trilithons which I have labeled in Fig. 12 "Moon" both align with two sarsen circle arches to make four alignments.

In addition to placing their huge stones in precise astronomic alignment, the last builders also placed them in such a way as to leave undisturbed most of the existing alignments, even though those alignments were duplicated in their stones. They chose for their sarsen circle radius a distance such that the northeast and southwest stones of that circle just missed, by a few feet only, intersecting the old 91–94 and 92–93 viewing lines.‡ Of the 16 alignments of Stonehenge I and II, all but five—center-91, center-93, center-A, center-D and 93-H—were preserved when the inner circles and horseshoes of Stonehenge III were added. Very artfully they maintained and duplicated orientations of a rectangular configuration—the Stonehenge I-II axis and the station stones—in a double-curved configuration, the Stonehenge III sarsen circle and trilithon horseshoe.

That final megalithic temple to the sun and moon required of its creators an absolutely extraordinary blending of theoretical, planning abilities with practical building skills. Consider the problem they set for themselves: to design and erect a circle enclosing a horseshoe in such a way that the units of both figures were regularly spaced and yet so arranged that the 5 narrow archways of the horseshoe aligned with 7 narrow archways of the circle to point to 7 of the 12 unique sun and moon horizon positions while the axis of the whole structure pointed through another circle archway to an eighth celestial position—all this to be managed with primitive tools, using "units" of stone, gigantic blocks weighing 30 tons or more. How well they solved that problem we see today.

The first builders—or rather we should say single designers with their groups of builders, because obviously there was directed planning before the construction gangs started work—needed intelligence, purpose and patience as well as physical skill and strength to create Stonehenge I. For Stonehenge II, more intelligence, and continuing purpose were required. To complete the great structure, incorporating the earlier works into a unified whole, a monumental temple with intricate celestial alignments concealed in apparent simplicity and

‡ This fact, that the sarsen circle circumference falls just within the station stone rectangle, has long been noted—and set aside as one of the meaningless coincidences or insoluble mysteries of Stonehenge. The discovery of the solar-lunar alignments makes it obvious, I think, that the Stonehenge III builders who designed their own alignments with such care were equally careful not to disturb the older ones; they laid out their largest stone circle with regard for dimension as well as orientation.

symmetry of design—that required intelligence of a still higher order, a single purpose steadfastly maintained during three hundred years of changing populations, customs, and cultures, and varied skills beyond those possessed by many twentieth-century men.

Look again at the diagrams.

Notice the economy of design, the use of one position in more than one alignment.

Of the pairings of Stonehenge I, 8 point to the sun and 8 to the moon; the total is 16 paired alignments. Yet instead of 32 positions, only 11 positions are involved. All of the special stones, and the center, were used in these alignments, 6 of the positions more than once, 2 of them 6 times each.

Now look at the lines of Stonehenge III. There are four more sun pointers, and four more moon alignments obtained by use of each of the "moon" trilithons twice. And here let me emphasize that these trilithon-sarsen circle archway viewing lines have not been capriciously chosen from a plethora of possibilities, to fit the astronomy. If you stand in that horseshoe, as I did, and try to look through the trilithon archways down viewing lines other than those shown on the diagram, you will find, as I did, that you cannot. Your view is constricted by the narrowness of the archways. You cannot look down lines which would point to no meaningful sun or moon position; you are forced to look through paired archways toward those inevitable sun-moon extreme positions. What is more, those hollows in the trilithon uprights—earlier mentioned and commonly supposed to have been caused by weathering—make possible the side-angled views. I think those hollows were not caused by weathering; I think they were deliberately bashed out of the stones to make room for the viewer's head.

To sum up, then: Stonehenge I had 11 key positions, every one of which paired with another, often more than one other, to point 16 times to ten of the twelve extremes of the sun or moon; Stonehenge III with its five trilithons and heel stone axis pointed 8 times to eight of those same extremes.

Such correlation could not have been coincidental.

Once the machine had established that the Stonehenge builders had aligned their monument-temple to the sun and moon with such skill and persistence and impressiveness, the question of course arose, Why? Why had they gone to all that trouble?

As I noted in *Splendor in the Sky*, two stones are all that is necessary to mark sunrise, or any other celestial point—why had the

Stonehengers taken such tremendous pains over their many alignments?

Only the archaeologists and other students of the past can ever answer that question. We astronomers with our computing machines can only provide facts for the trained fancies of those ancient-man specialists to play over.

But I would like to put forward this opinion.

The Stonehenge sun-moon alignments were created and elaborated for two, possibly three, reasons: they made a calendar, particularly useful to tell the time for planting crops; they helped to create and maintain priestly power, by enabling the priest to call out the multitude to see the spectacular risings and settings of the sun and moon, most especially the midsummer sunrise over the heel stone and midwinter sunset through the great trilithon, and possibly they served as an intellectual game.

To amplify a little on those three supposed reasons, let me state that it is well known that methods for determining the times of planting were of most vital concern to primitive men. Those times are hard to detect. One can't count backwards from the fine warm days, one must use some other means. And what better means could there be for following the seasons than observation of those most regular and predictable recurring objects, the heavenly bodies? Even in classic times there were still elaborate sets of instructions to help farmers to time their planting by celestial phenomena. Discussing the "deepe question" of the "fit time and season of sowing corne," Pliny declared, "this would bee handled and considered upon with exceeding great care and regard; as depending for the most part of Astronomie. . . ." Doubtless there are today many farmers who time their planting by the sky.

As for the value of Stonehenge as a priestly power-enhancer, it seems quite possible that the man who could call the people to see the god of day or night appear or disappear between those mighty arches and over that distant horizon would attract to himself some of the aura of deity. Indeed, the whole people who possessed such a monument and temple must have felt lifted up.

The other possible reason for the astronomical ingenuity and contrivance of Stonehenge is, I must admit, my own invention. I think that those Stonehengers were true ancestors of ours. I think that the men who designed its various parts, and perhaps even some of the men who helped to build those parts, enjoyed the mental exercise above and beyond the call of duty. I think that when they had solved the problem of the alignments efficiently but unspectacularly, as they had in Stonehenge I, they couldn't let the matter rest. They had to set themselves more challenges, and try for more difficult, rewarding, and spectacular solutions, partly for the greater glory of God, but partly for the joy of man, the thinking animal. I wonder if some day some authority will establish a connection between the spirit which animated the Stonehenge builders and that which inspired the creators of the Parthenon, and the Gothic cathedrals, and the first space craft to go to Mars.

In any case, for whatever reasons those Stonehenge builders built as they did, their final, completed creation was a marvel. As intricately aligned as an interlocking series of astronomical observing instruments (which indeed it was) and yet architecturally perfectly simple, in function subtle and elaborate, in appearance stark, imposing, awesome, Stonehenge was a thing of surpassing ingenuity of design, variety of usefulness and grandeur—in concept and construction an eighth wonder of the ancient world.

The seven classic wonders of the world were the pyramids, as a group, (or the Great Pyramid), the Hanging Gardens of Babylon, the statue of Zeus at Olympia, the temple of Diana at Ephesus, the mausoleum at Halicarnassus, the Colossus of Rhodes, and the Pharos lighthouse, at Alexandria. With the exception of the more perishable parts of the Babylonian gardens and the colossus—supposedly a 280-foot figure of brass—all of those wonders would seem to have been of stone. Yet surely in none of them was stone itself so skillfully used to record the fruits of intellectual endeavor in an emotion-inspiring temple as in the great monument on Salisbury Plain.

Chapter 8

THE RESPONSE

I sent a report of the Stonehenge findings to the British scientific journal *Nature,* counterpart of the American publication *Science.* "Stonehenge Decoded," which is reprinted in the appendix of this book, was published in *Nature* on October 26, 1963. The response to that article was immediate.

The London *Times* commented on the report on the day of its publication. The *Times* story was very good. It was accurate, clear, and ended with the observation:

"Professor Hawkins . . . may not himself carry archaeologists the whole way with his arguments, but [he] has given them more to bite on than they have had before from any astronomer."

That statement neatly summed up the indicated consequence of the machine's findings mentioned at the end of the last chapter: astronomy had established that there were many sun-moon alignments at Stonehenge—archaeology should seek to determine why.

A general response to my article followed soon. It was spirited, and astonishingly voluminous.

Stonehenge had interested me for only ten years, although I was born in England and had visited the site often. I now found that the old monument, or the idea of it, has intrigued people who have never been near it.

As a Pennsylvanian put it, "The massive character and the gradual dispelling of medieval superstitions about this monument necessarily fire the imagination and curiosity of anyone who takes an armchair interest in the work of the archaeologists and prehistorians. . . ." Another man called himself an "amateur 'student' of Stonehenge," and noted, rather typically, "I have read most that I could obtain regarding the site." And a California couple wrote, "We are fascinated by your evidence of the amazing skill of those long-gone people."

Letters came from all kinds of writers, from many countries—

Australia, Norway, France, Belgium, Sweden, Chile, the United States, Denmark, Holland, Uganda, Germany, Scotland. . . . They are still coming. It is heartening to think that there can be such concern—you might almost call it affection—for something so innocent of profit, pride and prejudice.

I must say, however, that the vigorous response to the Stonehenge article took me by surprise. This was my sixty-first scientific paper, and many of the others have seemed to me more exciting.

For instance, in 1963 I published an article on tektites, those weird spatterings of once-molten glass which are found at widely separated places on the earth. Tektites are fascinating things, and quite mysterious—it is not known how they were formed, or where. Some researchers believe they came from space, others think that they were formed here on earth, a result of materials being melted by the impact of giant meteorites. The presently favored theory is that they were formed by meteorites hitting the moon and melting its surface material into glassy blobs, some of which were jarred loose with enough velocity to escape the moon's gravity and fall to earth. I wrote (and continue to believe) that tektites were formed on earth.

I also once published a new theory of the universe, postulating a static cosmology in which there is continual use and re-use of matter and energy, and thus an eternity of existence. This theory is in conflict with the presently popular evolving universe hypothesis, which predicts an ultimate end for the universe; it cannot be proved or disproved until further astronomic measurements are made.

Most of my other papers have dealt with those pragmatically important space wanderers, the numerous and fast-flying meteors. For years there has been a laborious mapping of the paths of these particles, which range in size from smaller-than-pinhead to larger-than-locomotive, and move at speeds up to 60,000 miles per hour. This mapping has not been for entirely academic purposes. A meteor could easily fly right through a space vehicle and its occupants. It is good to know where these "space vermin" are most frequent. Meteor frequency also affects radio wave reflection, and the seeding of rain clouds by meteor dust.

But no scientific article that I have ever written has caused a general stir to compare at all with the commotion aroused by "Stonehenge Decoded." I am still a little surprised, and puzzled. Was it possibly because of the unusual juxtaposition of new and old —the use of the most modern, impersonal device, a machine, to look for human secrets hidden in stones older than history? If so, the interest was accurately focused, because that juxtaposition was almost physical: in this operation I had been little more than a mid-dleman, a means of bringing machine to monument . . . or vice versa.

Most of those early communications—postcards, letters, even an occasional telegram—were simple requests for reprints of the *Nature* article. (My supply of reprints was quickly exhausted and has had to be replenished several times since.) Some were longer, and contained comments, criticisms and suggestions. The spectrum was wide.

The comments often concerned the writer's own theories and beliefs, some of them quite intriguing, some bizarre. For instance, from Spain came a little booklet purporting to prove that the "Taulas" of Minorca, some eighteen megalithic monuments, were oriented to the sun and moon. The energy and intelligence that had gone into the creation of those "Templos Astrales," or astro-temples, must have guaranteed "un gran esplendor de las costumbres rituales," a great splendor of ritual customs, the author declared. A very inter-esting theory—but one which I personally cannot now check. I wish I could; I wish that many, or all, of the hundreds of Neolithic and Bronze Age stone monuments still standing could be accurately surveyed and examined for astronomic orientation. Much information fruitful for archaeologists, anthropologists, historians and others might result. If any university or foundation is casting about for promising fields for exploration and research, let it consider astro-archaeology!

"A student of myth," wrote a New Yorker, "learns early that religion and the calendar are the same thing in the young history of men and that temples were observatories and laboratories. Hence I was grateful for proof of the inevitable nature of Stonehenge."

A Massachusetts lady advanced her theory that the Stonehenge alignment errors were "possibly . . . deliberate," for the reason that "primitive people often do not make their work exactly perfect be-cause of their belief that only God makes perfect things."

A Californian wondered if Stonehenge "might possibly be many centuries older. Forgetting for the moment the length of time intel-ligent beings are presumed to have occupied the earth, would your calculations show a lesser error if the time cycle were moved back in units of as much as 25,000 years? I picked 25,000 years as a unit because I have a vague recollection that that is approximately the time it takes the solar system to go through one 'revolution' to bring certain relationships back into relatively the same position."

Answer: Stonehenge couldn't have been built in 25,000 B.C. for several reasons, one of which was ice—England was under it then.

But the 25,000-year solar system "revolution" was almost precisely recollected by this writer. The sun's "great year" is 26,000 earth years. In that period the sun as seen from earth slips back one revolution around its path, the ecliptic. Most people think that the sun goes around the ecliptic every year, and it does—almost. It doesn't go *all* the way around. It falls short of a compete 360° circuit by a little less than 1', or 1/60 of a degree, so that in 26,000 years it slides backward along the ecliptic one full revolution. That is why the relation changes between the twelve months and the twelve constellations of the Zodiac which lie along the sun's path through the heavens.*

None of the people who wrote to me mentioned the computer itself, either favorably or otherwise. But some of them did discourse on matters not confined to astronomy.

Thus, from Indiana came this letter: "You say in question, 'why is the heel stone ever so slightly out of line . . . ?' Maybe it wasn't out of line when Stonehenge was in use? . . . Maybe the difference measures a shift in land position since that time?" To that questioner, who suspected that she might be "romancing geologically," I had to reply that she was—there is no evidence that the land at Stonehenge has ever split open. The stones and holes are very probably just as they were in 1500 B.C.

And from England: . . . "The *numerical structure* and the *geometry* of Stonehenge . . . gave a *preview* of *Christianity*. It monumentalises the date of the Nativity, Crucifixion, Baptism, etc.—and focusses upon the present . . . it is often referred to in the Bible as 'Jerusalem' or 'Zion'—especially in Psalm 48:12 and in Daniel 9:25 (and in Psalm 122). It certainly is a *divine chronometer* in more ways than one." (Psalm 48:12 is an order—"Walk about Zion, and go round about her: tell the towers thereof." Daniel 9:25 says that the rebuilding of Jerusalem took "seven weeks, and threescore and two weeks." Psalm 122 speaks of Jerusalem, "a city that is compact together.") An Australian summed it all up briskly enough

---

* This changing relation has of course always been of utmost concern to astrology, although few of its present practitioners understand the astronomic cause. Few of its present practitioners understand *any* astronomy—but this is not the place to discourse on the strange career of astrology, that currently popular, highly profitable and utterly illogical business which Kepler called "the foolish little daughter which must sell herself to every bidder in order that her wise mother Astronomy should be able to live." Actually the foolish little daughter may be as old as the wise mother, going back to the time when men believed that souls came physically from the heavens as meteors, or "shooting stars," to inspirit unborn children; naturally, each descending soul was thought to be directly affected by the characters of the stars and constellations through which it passed. Then, astrology was a serious and even noble study. Now . . . well, *chacun à son goût.*

with this pronouncement: "Everything at Stonehenge is a sexual symbol."

I am told, however, that the number of odd responses in my mail was unusually small. Connoisseurs of the curious say that the appearance of any story to do with any aspect of science vs. mystery almost automatically releases a large impassioned "fringe" response these days. Whatever the subject of the triggering article, these responders are said to bring the discussion quickly around to their particular chosen realm of conjecture: Atlantis, or the equally lost continent of Mu; who-wrote-Shakespeare?; flying saucers; the Abominable Snowman; radioactive monsters.

As the letter writers busied themselves, so did the newspaper and magazine reporters. There was world-wide coverage of the Stonehenge story; journals from South Africa to North Carolina carried articles and editorials. Within one three-week period there appeared Stonehenge stories in the *Iraq Times* of Bagdad, and the Jerusalem *Post*. I found that the Arabic for "big stones" is "Hijarat Kabira," حِجارة كبيرة and the Hebrew is "Avanim g'dolim," אבנים גדולים. And in Jerusalem, 1500 B.C. is 1500 B.C.E., "before the Christian era."

Somehow, "El Misterio de Stonehenge," as *El Noticiero Universal de Barcelona* phrased it, looks quietly mysterious in foreign languages, except maybe German: "Rätsel Stonehenge Gelöst," The *Welt am Sonntag* of Hamburg seemed to shout, meaning, however, nothing more explosive than "Puzzle of Stonehenge Solved."

The newspaper and magazine stories were, in the main, commendably accurate. The *New York Times* published an extremely accurate and comprehensive report, and the *Manchester Guardian* was particularly sapient in its interpretation of my article. Remarking that the computer's finding "is bound to fire the archaeologists with fresh enthusiasm, and the Ministry of Works is going to find it harder than ever to keep the turf around Stonehenge intact"—which may or may not be—this paper pounced on the civic implications of the construction, implications which I have discussed earlier in this book. Nowadays, the *Guardian* continued, some Englishmen fear the "heavy burden" of research—"but has any project of civil research ever imposed so great a burden on the human resources of its day as the first research project of all in this country—Stonehenge?" The editorial emphasized the "care and time it must have taken to construct a pattern of stones so complex that its full significance has only been shown up by an electronic computer," and concluded, "Descriptions of Stonehenge commonly touch on the difficulty of moving the stones to the site. But knowing where to put them must have been much more difficult, and put the greatest strain on the scientific manpower of the day."

Not all of the newspapers were rigorously accurate in their re-reporting of the *Nature* report, though.

Two of them got tangled among the stars and planets. Although my article had made it *very* clear—I thought—that the machine had found *no* significant correlation between the Stonehenge alignments and the stars, or the planets, one paper wrote, "The data included correlations between the directions to find the lines joining various stones and holes, and the directions of the rising and setting of the sun and moon, as well as the movement of stars and other planets [sic] at midsummer and midwinter during the ancient era." The other paper took a geometrically clearer if astronomically murkier position: "Angles computed from diagonals drawn between key stones accurately describe the movement of stars and planets at the estimated origin of Stonehenge, within one degree."

An Ohio paper got all stones and pits lined up in concentric circles, which is untrue for the two horseshoes, the heel stone, the station stones and the other outlying stones, and then remarked mysteriously that at the center of all those alleged concentric circles was a "grass aisle." A Massachusetts paper set some sort of a record for compaction of errors with this one sentence: "They [the Stonehenge stones] are believed to have been erected by a tribe of ancient Druids some 500 years before the building of the Great Pyramid of Egypt." The facts are, as I have pointed out, that the druids very probably did *not* build Stonehenge and indeed may not have been in existence as a group when it was built, and Stonehenge was built nearly 1000 years *after* the Great Pyramid. That paper went on to brighten the whole field of archaeology greatly, by transposing two letters to produce the opinion that Stonehenge might have been a "marital court of justice."

A highly respected New York newspaper made all of these strange statements in one story:

> ". . . A secret 3000 years old . . . 1500 BC, the approximate year when Stonehenge is believed to have been built . . . the heel stone, at the junction of the avenue and the ditch . . . Aubrey holes . . . named for James Aubrey . . . a series of stone columns, called the Sarsen trilithons, and an inner horseshoe of 40-ton blocks . . . axis of the avenue proved to be the 'line of best fit' between summer sunrise and sunset. . . ."

In that welter of reportage are no less than six (6) full-grown errors. Item—the secret is at least 3500 years old. Item—Stonehenge was not built in any single year. Item—the heel stone is not at the Avenue-ditch junction. Item—Aubrey was John not James. Item—the "stone columns" were in the sarsen circle; the "40-ton blocks" in the "inner horseshoe" were the trilithon uprights. Item—the Avenue axis was of course the line of best fit between *midsummer* sunrise and *midwinter* sunset.

As a matter of fact, a popular national news magazine garbled the story almost as thoroughly as did that New York paper. This magazine's story started by defining the word "Stonehenge" as meaning "upright stones," although the Old English root "henge" means "hanging." Then, declared this magazine, I first used the computer to calculate the sun-moon rise—set points at midsummer and midwinter, "and their highest and lowest positions in the sky." Not so, of course. The first task of the computer was to calculate paired position horizon declinations and the machine was never used to calculate sun-moon "highest and lowest positions in the sky," whatever those, particularly the lowest, might be. "Then," continued this story, I "instructed the computer to work out all the varied angles among the stones and pits. . . ." "All the varied angles among" more than a hundred positions is a lot of "varied" angles indeed. Next, the magazine informed its readers, "Hawkins found that the sighting angles from stone to stone corresponded with remarkable accuracy to fourteen different key positions of the moon and ten of the sun." I don't think even the astrologers recognize that many "key positions" of the moon and the sun.

But such inaccuracies are excusable; the astronomic alignments at Stonehenge are not easily understood quickly; and in the main, as I have said, the press coverage of the story was very good.

The magazine *Holiday*, February, 1964, interpreted—or decoded— "Stonehenge Decoded" accurately and with great good humor: "The site has called up visions of gore and grue, of weird Druidic mysteries, of chilling rites in a prehistoric setting. The fantasy was natural! Whoever took the trouble to spend 500 years lugging those huge stones 200 miles from southwest Wales to the English plains near Salisbury, must have been driven by something sinister. . . ." *Holiday* conceded that "if Dr. Hawkins is right, another chunk of lore will have to be cashed in for the newer currency of fact," then comforted its holiday-minded readers, "But in a way this makes Stonehenge even more fascinating, and the site is easy enough to reach. It is two hours southwest of London, over good roads . . . ," and concluded, "the best spot at Stonehenge to ponder Dr. Hawkins' theory is, ideally, somewhere near the altar stone. The stone is probably mis-named, the purpose of the monument probably misunderstood. While night falls over the gaunt plain, silhouetting the old sundial's giant slabs, you can tell yourself this and wonder what ghastly purpose our descendants 3,000 years from now will attribute to the I.B.M. contraption."

Holiday elsewhere referred to the computer as "an I.B.M. know-it-all," and I thought that such levity of attitude toward their product might offend I.B.M.'s *amour propre*—until I saw an article from the *I.B.M. News* itself. This article began, "Those crazy old druids may have known what they were doing after all." The *News* paid scant attention to the computer—which "Hawkins . . . used . . . to help substantiate his theory . . ."—and dwelt more on its own theory that "those crazy old druids . . . labored mightily to set up massive stones and dig pits. . . ."

"Stonehenge Decoded" became lyrical when it was made the subject of a poem in *The Christian Science Monitor*:

STONEHENGE*

(Computer Finds Stonehenge Clues—A Headline)

Circle of stone,
you put a pedometer on the sun
and timed the moon.

Computer rocks,
giant monoliths
made a calendar of daily span.

The heavens whirl
virtually in the groove
they chose four thousand years ago:

Huge sandstone leans
three inches pushed aside
by centuries' blundering.

But these blunt digits
finger a fugitive sun
and build a cagey cage to catch its light.

CAROL EARLE CHAPIN

* Reprinted by permission from *The Christian Science Monitor*, © 1963 The Christian Science Publishing Society. All rights reserved.

My scientific article in *Nature* even made the funny papers. On September 13, 1964, a syndicated Sunday comic page strip, "Our New Stone Age" by astronomer Athelstan Spilhaus, presented handsome pictures showing a well-dressed man leaning pensively against a sarsen circle upright; another man in a laboratory, presumably carbon-dating; a yellow-haired youngster, presumably me, at a computer control panel, "correlating alignments of nearly 200 pairs of stones with the rising and setting of heavenly bodies as of 2000–1500 B.C."; and, finally, stones silhouetted against the twilight, with this conclusion: "the results showed that various pairs of stones pointed to the most northern and most southern points of rising and setting of the sun and moon. Stonehenge was an accurate astronomical observatory!" That funny paper panel was a factual report, except that I am not all that young, and my hair is black.

Among all the responses to "Stonehenge Decoded" there were two which were of exceptional interest. Both were from qualified archaeologists; one was adversely critical, the other was guardedly favorable, and was to prove extremely helpful in directing my attention to other secrets of Stonehenge.

Monsieur G. Charrière of France attacked my conclusion that the solar-lunar alignments were significant on the grounds that circles are "undifferentiated" axially whereas I had assumed otherwise; that I had "arbitrarily" designated certain stones as more significant than others, and that I had derived my sun-moon declinations on the "entirely personal notion" that the Stonehengers had used the moment when the disc of the sun or moon stands tangent on the horizon to determine the horizon point of rise and set.

I have written to M. Charrière amplifying the account of my research which appears in this book, pointing out that in my opinion the Stonehenge circles which were used in the research were not undifferentiated axially but were oriented by major axes to the midsummer sunrise; that, as the charts show, the stones and positions used in the correlations were all in some way unique or special, and that the tangent-on-the-horizon position of the disc was the one that gave the smallest error for 1800 B.C. midsummer sunrise over the heel stone, an alignment universally conceded to have been intended by the builders. I can only hope that further correspondence will resolve these differences of interpretation.

The other, most fruitful, letter came from R. S. Newall, the British archaeologist who has taken part in excavations at Stonehenge and is the author of the official guidebook. Writing from his home near the site, he stated:

"It is always difficult, I suppose, when two different sciences meet (if archaeology can be called a science), to come to agreement. Astronomers have their eyes in the sky; archaeologists in the earth . . . however I agree that Stonehenge is oriented to the winter solstice setting sun in the great central trilithon as seen from the center or anywhere else on the axis, and since the plan of Stonehenge is sepulchral it is in some way the mortuary temple to the sun in his old age when he goes down to the lower world at the end of the year or life . . . the heelstone since it is nearly on the axis line must necessarily be in line with the summer solstice sun rise and I have no doubt it was the whole orb that was observed, as in Egypt. This applies to the moon too. . . .

"Mr. Newham points out that a line 94 to stone hole C on the Avenue is the equinox sun rise, i.e. due East. Would that have been so in 1500–1000 B.C. or is it a coincidence? . . . [If it was not a coincidence] then the man who placed those stones . . . must have been a Genius. . . .

"Another point of interest is the Greek author Diodorus [who] . . . mentions . . . this temple in the land of the Celts which is . . . 'spherical in shape' . . . can 'spherical in shape' mean 'spherical in use,' i.e. astronomical? If so then somewhere in the land of the Celts at some time there was an astronomical temple. He [Diodorus] says 'the god [Apollo, the sun god] visits the island [presumably England] every nineteen years . . . he plays the cithara and dances the night through from the vernal equinox until the rising of the Pleiades.' Now I do *not* say that that refers to Stonehenge. But could it . . . ? Could the full moon do something spectacular once every nineteen years at Stonehenge? If it did, well I would not know what to say."

The "Mr. Newham" referred to is C. A. Newham, a keen student of the astronomy and geometry of Stonehenge.† His cited statement about equinoctial alignment of the station stone position 94 and stone hole C was thought-provoking.

As for Diodorus of Sicily, the so-called "universal historian" of the first century B.C., Stonehenge literature abounds in references to his account of Apollo's temple in the land of the Hyperboreans. The "happy Hyperboreans" appear often in classic writings, usually as a fortunate people living in the far north, "beyond the north wind," who worshiped the sun god Apollo. It is probable that they were the

† Newham has done much good work in measurement and survey at Stonehenge and I must record my gratitude to him for providing me with skyline altitudes.

real inhabitants of northern lands, imaginatively described by the travelers and traders, particularly the amber traders accustomed to going to the Baltic, and given more mythical traits partly because of the mythical sound of their habitation. To the Mediterraneans any people who willfully lived so far from the sun must have seemed mad or mythical.

The poet Aristeas placed them next to the legendary "one-eyed Arimaspi" and "gold-guarding Griffins," but Herodotus made them sound un-legendary enough. He reported that the people of Delos said that the Hyperboreans sent "certain offerings, packed in wheaten straw," all the way from their northern land to Delos, Apollo's island. Since the northerners had once sent girls who had not returned, they cautiously sent their straw-wrapped offerings by human chain, trusting them to be handed on from country to country and city to city until they reached their destination. Herodotus said that there were still customs among the people of Delos stemming back to the honoring of four maidens who had come there from Hyperborea, and concluded, "As for the tale of Abaris, who is said to have been a Hyperborean, and to have gone with his arrow all round the world without once eating, I shall pass it by in silence."

Pliny described the Hyperboreans thus: "For six moneths together they have one entire day, and night as long . . . the countrey is . . . of a blissful and pleasant temperature . . . their habitations be in woods and groves, where they worship the gods . . . no discord know they; no sicknesse . . . they never die, but when they have lived long enough: for when the aged men have made good cheere, and annointed their bodies with sweet ointments, they leape from off a certain rocke into the sea . . . in the nights [they] lye close shut up within caves. . . ."

Diodorus‡ gave this account, which has intrigued Stonehenge students from Gidley and John Wood on:

"This island . . . is situated in the north, and is inhabited by the Hyperboreans, who are called by that name because their home is beyond the point whence the north wind (Boreas) blows; and the land is both fertile and productive of every crop, and since it has an unusually temperate climate it produces two harvests each year. Moreover, the following legend is told concerning it: Leto [mother of Apollo and Artemis—Zeus was their father] was born on this island, and for that reason Apollo is honoured among them

‡ Book II, Loeb Library translation.

above all other gods; and the inhabitants are looked upon as priests of Apollo, after a manner, since daily they praise this god continuously in song and honour him exceedingly. And there is also on the island both a magnificent sacred precinct of Apollo and a notable temple which is adorned with many votive offerings and is spherical in shape. Furthermore, a city is there which is sacred to this god, and the majority of its inhabitants are players on the cithara; and these continually play on this instrument in the temple and sing hymns of praise to the god, glorifying his deeds.

"The Hyperboreans also have a language . . . peculiar to them, and are most friendly disposed towards the Greeks, and especially towards the Athenians and the Delians, who have inherited this goodwill from most ancient times. The myth also relates that certain Greeks visited the Hyperboreans and left behind them there costly votive offerings bearing inscriptions in Greek letters. And in the same way Abaris, a Hyperborean, came to Greece in ancient times and renewed the good-will and kinship of his people to the Delians. They say also that the moon, as viewed from this island, appears to be but a little distance from the earth and to have upon it prominences, like those of the earth, which are visible to the eye. The account is also given that the god visits the island every nineteen years, the period in which the return of the stars to the same place in the heavens is accomplished; and for this reason the nineteen-year period is called by the Greeks the 'year of Meton.'§ At the time of this appearance of the god he both plays on the cithara and dances continuously the night through from the vernal equinox until the rising of the Pleiades, expressing in this manner his delight in his successes. And the kings of this city and the supervisors of the sacred precinct are called Boreades, since they are descendants of Boreas, and the succession to these positions is always kept in their family."

Diodorus elsewhere (Books IV, III) discussed astronomy, declaring that Atlas "discovered the spherical nature of the stars," and perfected the "science of astronolgy . . . and it was for this reason that the idea was held that the entire heavens were supported upon the shoulders of Atlas, the myth darkly hinting in this way at his discovery and description of the sphere."

To one interested in possible astronomical aspects of Stonehenge

§ The fifth century B.C. Greek astronomer Meton noted that 235 lunar months equal 19 solar years, so that after one "metonic cycle" of 19 years the full moon occurs again on the same calendar date.

PLATE 13. During World War II Stonehenge was used to test the feasibility of flash photography from a moving aircraft for reconnaissance behind the enemy lines. With uncanny precision the strobe flash from the aircraft at 5000 feet has been set off exactly over the ring of stones. Dr. Harold E. Edgerton, who took this photograph, was a pioneer of research in stroboscopic photography, the foundation for the development of the present-day electronic speed flash.

PLATE 14. The sunrise trilithon 51–52, and the view through the sarsen archway 6–7.

PLATE 15. The moonset trilithon 57–58, and the view through the sarsen arch 21–22.

PLATE 16. The prehistoric "stage set," showing the heel stone framed in the archway 30-1.

PLATE 17. A wide-angle camera view of sunrise, June 20, 1964.

PLATE 18. A telephoto camera view of sunrise over the heel stone June 20, 1964. In 2000 B.C. the sun would have been one diameter higher.

PLATE 19. A few minutes after sunrise, June 19, 1955.

PLATE 20. Midsummer sunrise viewed along the axis of the monument, through the remaining half of the great trilithon, through sarsen 30–1 and looking toward the distant heel stone, June 22, 1962.

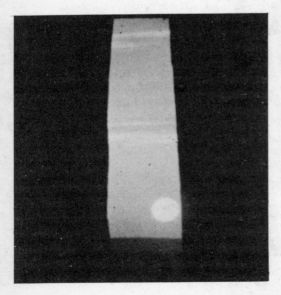

PLATE 21. A few minutes after the midwinter sun had risen. The camera was placed near the fallen stone 59, and this telephoto picture was taken through the narrow gap in the sunrise trilithon 51–52 and the sarsen archway 6–7. In 2000 B.C. the sun would have been one diameter closer to the horizon. December 1964.

PLATE 22. A wide-angle view of the midwinter sun in the sunrise trilithon, December 1964.

PLATE 23. The midwinter sun setting in the approximate direction of the great trilithon, December 1964. Stone 56 is standing, but its companion, 55, to the left, has fallen.

Diodorus' account can be richly suggestive. He reported, in a matter-of-fact style, that in a northern island there was a spherical, or astronomic, temple to the sun god—a temple to which that god returned every nineteen years, "the period in which the return of the stars to the same place in the heavens is accomplished . . ."—and that the people of that island also were careful observers of the moon.

I was very grateful to Newall for calling my attention to those references of Diodorus, and also for mentioning the equally interesting matter of Newham's equinoctial alignment.

Reading that Newall letter, I began to feel that there might be further astronomical discoveries to be made at Stonehenge. The title of my article, "Stonehenge Decoded," seemed perhaps presumptuous and premature—Newham's equinox and Diodorus' nineteen years should be investigated.

More work should be done.

And so, inevitably, back once more to the machine.

Chapter 9

ECLIPSES

When at last in early 1964 I managed to return to the problem of Stonehenge and focused my attention on that matter of Newham's equinoctial alignment I experienced immediate strong embarrassment. I remembered then something about the first alignment readout which the machine had given us: along with the 29-, 24-, and 19-degree declinations which it had reported, those sun and moon extremes which we had investigated with such success, there had been two which we had *not* investigated—one near 0°, which marks the sun at equinox, and one near +5°, which can mark the midway moon.

I had noticed those two alignments. I had even speculated on the possibility that the +5° one (heel stone from 94) could have been meant to point to the rising of the Pleiades. Some Stonehenge authorities have advanced that theory. But I decided against it because in the first place the Pleiades then rose slightly north of the midway moon—about +6°43′ declination in 1750 B.C.—and in the second place six of those Seven Sisters are fourth-magnitude stars, too faint to be seen when rising, and the seventh is so dim that only a very bright-eyed observer can see it even under the best conditions.

I had suspected the near 0° declination, stones F-93, as a deliberate equinox sun alignment, but since no other alignment in that first readout produced a similar near-zero declination I had regarded F-93 as unconfirmed, and had seen no way to confirm it. We were then fitting a pattern of alignments to the north-south extremes of the summer-winter sun and moon, and did not concern ourselves with possible east-west alignments at the spring and fall equinoxes.

Newham's line provided us the clue. He had used stone hole C. We had omitted stone holes B, C, and E from our calculations, because they seemed to us to be nonunique. They lay so close to the center–heel-stone line that we figured they probably had been additional, rather clumsy markers in the alignment to midsummer sunrise. Since they seemed to have had no other use we did not consider them "key" enough to be put into the machine.

I went back to the machine in January, 1964—two years after the first calculation—and gave it these additional positions, B, C, E. Again its report was astonishing. (Table 2.) Stone holes B, C, E, F aligned with station stone positions 93 and 94 to produce four near-zero declinations, close to the sun's equinox position, and four near-5° declinations, three north and one south, similarly close to two of the moon's four midway points. (Fig. 14.)

TABLE 2

Position	Seen from	Azimuth Clockwise from North Degrees	Object and Declination Degrees		Distance Above or Below Skyline Degrees
Heel	94	82.7	Equinox moonrise	+5.2	−0.3
B	94	84.6	Equinox moonrise	+5.2	+1.0
F	93	89.0	Equinox sunrise	+0.0	−0.9
C	94	89.5	Equinox sunrise	+0.0	−0.5
E	94	100.1	Equinox moonrise	−5.2	+0.4
93	F	269.0	Equinox sunset	+0.0	+0.4
94	C	269.5	Equinox sunset	+0.0	+0.0
94	D	277.7	Equinox moonset	+5.2	−0.7

As might be expected, since it has two maxima, the moon does not always cross the halfway point in its north-south swingings at the celestial equator, declination 0°, as does the one-maximum sun. Because of those orbit plane motions discussed before, the full moon at the midway point can be anywhere from $5°15$ north to $5°15$ south of declination 0°. Whereas the sun will cross at declination 0° for as long as the earth endures, the moon may in some inconceivably distant future, change the limits of the midpoint swing from the present $5°15$—but the likelihood is slim. Therefore we did not have to chase the moon back to 1500 B.C. to check possible Stonehenge alignments at this stage of the calculations.

Those eight equinox or midpoint alignments were well within the

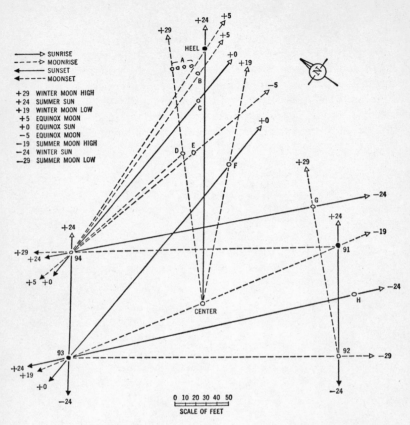

Fig. 14. All alignments found for Stonehenge I, including the equinox lines for the sun and moon.

limits of accuracy established for the 24 sun-moon extreme alignments discussed in Chapter 7.

I need hardly point out that this finding was of an importance comparable to that earlier one. The midway points are obviously significant. They are the halfway positions between the solar/lunar north-south extremes; just as the solstices mark the beginning of summer and winter, the solar equinoxes for us Machine Age men still officially mark the beginning of spring and fall.

Once the Stonehengers had got the solstices, or summer-winter extremes, aligned—what was more natural than that they would try for the midpoints? With equinoxes and solstices they could quarter the year.* They *could* have gotten these halfway points by bisecting the angles between solstice lines. Such a geometric method, familiar since long before Euclid, would have been easier than any observational technique. However it was done, the stones are aligned to equinoxes with remarkable accuracy.

Newham had been right. He was the first person to whom I sent Table 2. The machine had tried to tell us.

A consideration of the midpoint alignments inspired renewed respect for the Stonehenge builders. Again, as in the case of the extreme alignments, they had demonstrated skill in planning as well as accuracy of placement. The spring-fall alignments pointed to both of the sun positions and three of the four moon positions, with four duplications—yet to create these 8 paired alignments, not 16 but only 8 stones and holes had been used.

The addition of the equinox correlations meant that *every one* of the 14 key Stonehenge I positions was involved in at least one alignment with one of the 18 most significant celestial positions—those 14 Stonehenge positions were so placed that altogether they combined in pairs to produce 24 alignments—and Stonehenge III independently gave 8 more. Stonehenge was locked to the sun and moon as tightly as the tides.

Those astonishing figures fairly haunted me: 22 key earthly positions aligning, 32 times, with 15 of the 18 unique sun/moon positions. I had felt sure that the sun-moon extreme alignments found at first had been well beyond the possibility of coincidence. Now, the machine showed that all 14 key positions of Stonehenge I and all 8 "views" in Stonehenge III were involved in an elaborate network of sun-moon extreme or mean alignments. I wondered what the odds actually were against coincidence.

It was the standard problem of a blindfold marksman shooting at a target. And Bernoulli's law gives the answer: If the shooter has n shots and the target occupies a portion p of his shooting area, then the chance of scoring x hits is

---

* Pliny said (Bk. XVIII, Chap. 25), "Now all the knowledge of the heavens pertinent to Agriculture, standeth principally upon three sorts of observations, to wit, the rising of the fixed stars; the setting of the same; & the four cardinall points, to wit, of the two Tropickes or Sunsteads, and the double Aequinox, which divide the whole year into foure quarters and notable seasons."

$$\frac{n!}{x!\,(n-x)!}\ p^x(1-p)^{n-x}$$

Consider Stonehenge I: 14 stones and holes when paired together score 24 hits on the sun-moon targets, so x is 24. The diagram shows that the number of ways these positions can reasonably be paired is no more than 50, so let n be 50. How much of the 360° horizon is occupied by target area? There are 18 possible targets. We will define each target or bull's-eye to be 4° wide, so $p = 18 \times 4 \div 360$, or $\frac{1}{5}$.

These numbers put into Bernoulli's law would give the probability of scoring the 24 hits by luck alone, but the arithmetic would be horrible. (This probability calculation is based on the concept of randomness. Since Stonehenge obviously has a pattern, Bernoulli's law cannot be rigorously applied. If one is willing to admit that the Stonehengers were not "blindfold marksmen" then the probability model is invalid. But, then, the calculation to establish a non-chance pattern becomes less essential.) I personally let the machine do the figuring. The answer was 0.00006, which means less than one chance in ten thousand that the stones had been so aligned by accident.

Now consider Stonehenge III. Each one of the eight shots hits one of the sun-moon targets. Bernoulli's law shows that the chances are about a thousand to one against random alignment.

Stonehenge I and III are separate structures, and the odds against both structures having the alignments by accident are a thousand multiplied by ten thousand, or ten million to one—which is to say that the chance of Stonehenge being aligned to the sun and moon by coincidence is negligible.

Can more astronomically significant alignments of key positions be found at Stonehenge? Possibly not. As I have said, the machine has examined practically all of the alignments of indicated importance. Not unless new positions are found by excavation of the site and/or other means such as exploration of the surrounding area does there seem likelihood of discovery of more Stonehenge celestial correlation.

One can feel almost apologetic, in a way. Down through the centuries many good men have wondered about possible celestial alignments at Stonehenge. Long ago it was recognized that the main axis, the midsummer sunrise line, points almost exactly to midwinter sunset if extended in the reverse, southwest, direction. As far back as 1846 Duke noted that the station stone positions 92–91 parallel the axis or solstice line. Early in this century Lockyer showed that the station diagonal 91–93 marks sunset at about May 6 and August 8, and the reverse, 93–91, marks sunrise at about Feb. 7 and Nov. 8. Those days are approximately midway between solstices and equinoxes. He therefore suggested a calendar purpose. (An interesting suggestion, but one with which I disagree. I consider that this diagonal was intended to mark the moon at its maximum declinations ±19°, similar to the positions marked by the holes in the Avenue and by the moon trilithons. Admittedly the error of alignment is large for 91–93 in Stonehenge I, but this error is considerably reduced in Stonehenge III.)

Modern Stonehenge aficionados have done a good deal of speculating about possible significance, astronomical or otherwise, of alignments derivable at the monument. Newham himself has been particularly adept in noting possible celestial alignments. After I got that first letter from Newall quoting him, I started direct correspondence with Newham. It turned out that he too had been investigating some of the same Stonehenge-sun/moon alignments I had been busy with.

Newham published a brief account of his work in the *Yorkshire Post* on March 16, 1963—seven months before my *Nature* article appeared. (Needless to say, I knew nothing of his story when I wrote my article.) And he published a small booklet, *The Enigma of Stonehenge*, soon after "Stonehenge Decoded" was printed. In this booklet he was kind enough to acknowledge my article and my work. Since then we have established a most cordial relation and exchanged much information.

He had proposed solar/lunar alignment for 94-G, 92-G, 94–91 and 92–93. There can be little doubt that if he had been, like me, fortunate enough to have had the use of a computer he would have established the entire correlation. Indeed, I should here state once again that most of the credit for the solution to Stonehenge put forward in this book should go directly to the machine. That uncomplaining drudge in a few seconds performed the hundreds of messy calculations which for so long had discouraged would-be human investigators. I only hope that future students of Stonehenge, faced with whatever new problems the old monument may pose, will have the use of whatever descendants may have developed from the venerable 7090.

While working out those astronomic odds against the Stonehenge alignments being accidental, and trying to put a measure to the skill of those primitive astronomer-architect-engineer-workmen, I could not

dismiss from my mind the second of the questions which the archaeologist Newall had raised in the letter he had written in response to
my *Nature* article:

What was the meaning of Diodorus' nineteen years?

Of course, the number 19 is both ancient and common in astronomy. Diodorus himself mentioned the "metonic cycle," and certain
Jewish and Chinese calendars have used such a 19-year cycle. But
what had 19 to do with Stonehenge? Was it somehow visually connected with the moon?

As Newall himself had so succinctly phrased it, "Could the full
moon do something spectacular once every nineteen years at Stonehenge?"

Then suddenly I thought of the only really "spectacular" thing the
full moon can do—become eclipsed. I asked myself, "When is the
eclipsed moon most spectacular?" The answer came at once: "When
it is over the heel stone, or in the archway of the great trilithon."
The problem was becoming specific. We needed to investigate
eclipses.

Eclipses would clearly be among the most impressive and frightening natural phenomena that primitive men could encounter. What
terror would strike the people as the god, or goddess, was swallowed
up! Power and glory would surround the priest who could predict and
thus seem to control those monstrous events. And vice versa—the famous story of the Chinese court astronomers Hsi and Ho who missed
the solar eclipse of Oct. 22, 2137, B.C., and were promptly executed,
may not be entirely true, but personally I would not like to have been
the court astronomer of any country in any ancient time who failed
to warn of a coming eclipse.

Eclipse prediction is a venerable science, which doubtless was
made to appear an art, or a magical feat, by the initiated. Pliny wrote,

"True it is (I confesse) that the invention of the ephemerides
(to foreknow thereby not onely the day and night, with the eclypses
of Sun and Moone, but also the verie hours) is auncient: howbeit,
the most part of the common people have been and are of this
opinion (received by tradition from their forefathers) That all the
same is done by enchantments, & that by the means of some sor-
ceries and hearbes togither, both sun and moon may be charmed,
and enforced both to loose and recover their light: To do which
feat, women are thought to be more skilfull and meet than men.
And to say a truth, what a number of fabulous miracles are reported

to have been wrought by Medea queene of Colchis, and other women; and especially by Circe our famous witch here in Italy, who for her singular skill that way, was canonized a goddesse."†

Legend credits the Babylonians with eclipse predictions far back into antiquity, but a careful reading of the clay tables shows that they did not have much success until about 500 B.C. By then moon eclipses were calculated from the fact that the moon can be eclipsed only when it is full, and on the ecliptic. We will leave the problem of whether this fact was known in England 1000 years before the Babylonians as a moot question for the moment.

That fact, known since antiquity—that the moon must be just opposite the sun to be eclipsed—made our Machine Age task much simpler. Since without an impractically elaborate program we could not make the computer calculate the dates and positions of past eclipses, we instructed it to do the next best thing: calculate the positions of the full moon as it would have been seen from Stonehenge for every winter of the thousand years from 2000 to 1000 B.C. That task took the machine a few seconds only, and its report, arranged graphically, made an arresting pattern.

In a cycle of 18.61 years, the midwinter full moon moved from maximum north, declination +29° at stone D, across the heel stone to minimum north, declination +19° at stone F, and back again. Similarly the midsummer full moon moved back and forth across the viewing line through the archway of the great central trilithon.

Then I consulted the standard text on the subject of early eclipses, Van den Bergh's *Eclipses in the Second Millennium BC*, to find the months in which eclipses of sun or moon had taken place. The machine print-out then showed where the lunar eclipses had been.

The result was most instructive. It showed that an eclipse of the moon or the sun *always* occurred when the winter moon—that is, the full moon nearest the winter solstice—rose over the heel stone. Not more than half of those eclipses were visible from Stonehenge, but the good chance that the inevitable eclipse might have been visible from England would have made it well worth while for the Stonehenge priests to use winter moonrise over the heel stone as a danger signal. Far better to call the people out for a false alarm—and then perhaps claim that skilled intercession had averted the disaster—than

† Circe turned Odysseus' men into swine and detained the hero for a year; she and Medea and perhaps the Witch of Endor are the most famous of all witches. In classic times she was still feared, and it was thought that a certain tribe, the Marsi, were descended from her, and could therefore charm snakes.

to fail to call them out and have the eclipse come without warning!

Further work showed that when the swing of the winter moon carried it over D or F, then the harvest moon was eclipsed that year. The interval between the nights of winter moonrise over the extreme line center-D was about 19 years. But "about" is not "exactly." In this case "about 19" meant almost exactly 18.61—which meant that instead of the intervals between winter moonrises over D being a comfortable continuing series of metonic cycle 19 years, they came in a jumble of 19's and 18's, averaging two 19's to one 18 . . . which in turn meant that if the priests, intently tracking the years so as to be able to predict eclipse danger, had used a simple 19-year interval, they would have been right for perhaps two intervals, and then after a third would have been off by a full year. A rigid 19-year cycle would have soon drifted into hopeless error. The only regular-interval alternative, an 18-year cycle, would have been twice as bad. The smallest time unit that would have remained accurate for many years would have been the triple-interval measure, 19+19+18, or a total of 56 years. Our graphs showed that Stonehenge moon phenomena repeated every 56 years with good uniformity. The triple-interval of 56 years between winter moonrises over Stone D was accurate for centuries.

Therefore, we reasoned—trying to put ourselves back into the minds of Stone Age priests whose livelihoods and possibly lives might well have depended on eclipse prediction—he who would track the moon would use a 56-year cycle.

The figure 56 seemed familiar. It *was* familiar—it was one of the oldest, most puzzling mysteries of Stonehenge.

It was the number of Aubrey holes.

As was stated earlier in this book, there has never been a satisfactory, or even a tentative, solution advanced for the problem of the number of Aubrey holes. It has always been obvious that they were important: they were carefully spaced and deeply dug; they served, sporadically, the sacred purpose of tombs; filled with white chalk, they must have been compelling spectacles. But they never held stones, or posts—and, being so numerous and so evenly spaced, they could hardly have been useful as sighting points. What was their purpose?

I think that I have found the answer.

I believe that the 56 Aubrey holes served as a computer. By using them to count the years, the Stonehenge priests could have kept accurate track of the moon, and so have predicted danger periods for the most spectacular eclipses of the moon and the sun. In fact, the Aubrey circle could have been used to predict many celestial events.

It could have been done quite simply. If one stone was moved around the circle one position, or Aubrey hole, each year, all the extremes of the seasonal moon, and eclipses of the sun and moon at the solstices and equinoxes, could have been foreseen. If six stones, spaced 9, 9, 10, 9, 9, 10 Aubrey holes apart, were used, each of them moved one hole counterclockwise each year, astonishing power of prediction could have been achieved.

With six stones, three white, three black, the Aubrey hole computer could have predicted—quite accurately—every important moon event for hundreds of years.

The method could have been as follows.

Let us suppose the stones are placed as shown in Fig. 15,‡ and the year is 1554 B.C., the year of that appalling spectacle, a winter eclipse. The priests know of the danger of the winter eclipse because a white stone is at Aubrey hole 56. As confirmation of the danger period, and also as a check on the running of the computer, they watch to see the full moon rise over the heel stone; as it does they might say, "The winter moon has usurped the position of the summer sun—beware!" During the year when a white stone is at hole 56 the winter moon sets along the line G to 94. During such a year there would also be a second danger period for eclipses of the sun and moon, the month (i.e. the moon) of the summer solstice, when the full moon rises in the sunrise trilithon and sets in the great trilithon. The year 1554 would have given the priests a busy observing schedule —for which they would have been warned by that white stone at hole 56.

Next comes the year 1553 B.C. All stones are moved by one Aubrey hole in a counterclockwise direction. The white stone is at hole 55. This is a "safe" hole; nothing spectacular happens that year. The winter moon has swung part of the way over toward D, following the movement of the computer stone.

Actually nothing spectacular will happen for five years until the white stone is at Aubrey hole 51. Then what does the computer predict? It is the year 1549 B.C. The winter moon reaches its extreme declination of +29°. It rises over D-center, it sets along 94–91 and

‡ Here I am operating the Aubrey circle computer in a counterclockwise direction. In my *Nature* article (see appendix) I tried the clockwise rotation, for no particular reason (except, perhaps, that I am right-handed). Afterward a mathematician friend (who is left-handed) suggested that as a possible check I try it again, in reverse.

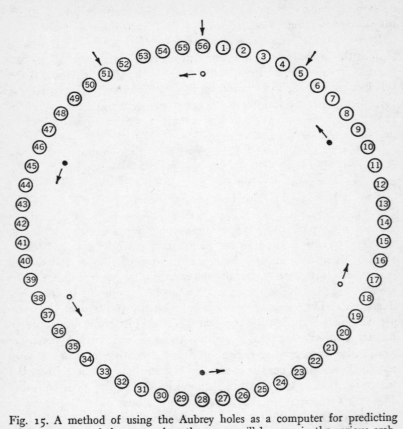

Fig. 15. A method of using the Aubrey holes as a computer for predicting eclipse seasons and the years when the moon will be seen in the various archways and stone alignments.

in the moonset trilithon. The summer moon rises along 92–93 and in the moonrise trilithon. The harvest moon and the spring moon rise and set along 94-C and 93-F. The danger periods for eclipses are the month of the harvest moon and the month of the spring moon, i.e., at the equinoxes. All this makes 1549 another busy year for the observer-priests, but comfortably expected because a white stone is at 51.

Four more years of safety pass by, then we come to 1545. A black stone is now at 56. All the moon events and eclipse dangers of 1554 are repeated—predicted by the occurrence of that computer stone at Aubrey hole 56.

In essence, a black or white stone at hole 56 occurs at intervals of

9, 9, 10, 9, 9, 10 years. This predicts the heel-stone moon events. A white stone is at hole 51 at intervals of 18, 19, 19 years, predicting the conditions of the high moon at +29°. A white stone is at hole 5 at intervals of 19, 19, 18 years, predicting the events associated with the low moon, at +19°.

As a modern illustration, let us use Stonehenge to fix the dates of Easter Sunday, the Passover, and all the associated religious observances.

When the sun rises and sets along 94-C and F-93, that day is the first day of spring. In the days that follow, the moon stone will move slot by slot around the sarsen circle as described in the *Nature* article in the appendix. When it arrives at the main archway, 30–1, that is the day of the spring full moon, which is the day of Passover. The following Sunday will be Easter Sunday.

When will the Easter moon, alias the Passover moon, alias the spring moon, be in danger of eclipse? Answer: when a white stone is at Aubrey hole 5 or 51. In that year the winter moon will be over D or F, the summer moon will be along the long sides of the station stone rectangle and the spring moon will rise along 94-C.

This modern use is not so fanciful as it might seem at first. Easter is linked to the Passover and the Passover has been traced far back to the fringes of prehistory. In ancient Continental Europe, there are many accounts of the festival of the fearsome "rites of spring" with sacrifice and fertility ritual. Easter comes from "Eostre," the Teutonic goddess of the spring.

The Easter egg is chocolate now, but in prehistory was a symbol of new life. The Easter bunny is an American version of the European Easter hare. Because the hare is born with its eyes open, in ancient times it was associated with the moon—"the open-eyed watcher of the skies." Our practice of wearing something new at Easter comes from an old custom—new fires were kindled by Teutonic tribes on this special day to mark the death of winter.

The date of Christmas has been arbitrarily established to replace the pagan midwinter festival. It is marked by sunset in the great trilithon—if the Stonehengers followed the customs of Continental Europe huge fires were then lit to signal the turning point of the sun. Lights on the Christmas tree are a vestige of this pagan ceremony. The Christmas moon is eclipsed when a black or white stone is at Aubrey hole 56, and so forth.

We are only 150 generations away from the European Stone Age. We have many customs, superstitions and perhaps even traits left over from prehistory. It is eerie but not really surprising to find that

Stonehenge could be put in motion to predict our modern movable feasts.

I have recently found that the Aubrey circle computer can be worked with three instead of six stones; the winter or summer eclipses occur when any stone is at hole 56 or 28, that is, on the axis. Actually, the predicting can be done with one stone only, if 12 positions are marked around the circle. This might be called the "Einstone" method.

Well, enough. I will not belabor the point. Anyone who is interested can use the diagram to work out more prediction powers and probabilities. I can assure him that there are a great many of them. As a computer, the Aubrey circle could have been a singularly effective instrument.§

Can it be proved that the Aubrey holes were used as a computer? Of course not. The situation is not parallel to the Stonehenge-sun/moon alignment. No law of probability can be invoked. Bernoulli's law does not apply.

All that can be said is that this proposed solution to the problem of why there were 56 Aubrey holes is the most reasonable one yet proposed. As a matter of fact, it is about the *only* solution that has been proposed.

In favor of this solution—that the Aubrey holes were used as a computer—are these facts: the number 56 is the smallest number that measures the swing of the moon with an over-all accuracy of better than 3 days, and lunar cycles provide the only method of long-range eclipse prediction related to the seasons of the year.¶ Stonehengers, like all primitive people, were probably concerned about eclipses, and they must have been particularly concerned about the moon—witness the alignment of 16 paired positions to unique moon declinations. They were capable of sustained, superior calculation, and engineering performance to match—witness the whole design and construction of Stonehenge. Existence of the 56-year lunar cycle could have been detected over a period of centuries—and the Stone

§ One day I happened to speak of this possible use of the Aubrey circle as a Stone Age computer to a Boston University research student who was well trained in computer technique, being a member of the new computer generation. He was not impressed. His scornful comment: "OK—so it was a computer—but it was only a single-purpose machine."

¶ The tropical year is about 365.25 days in length. That is, the instant of spring, the vernal equinox, occurs on the average after an interval of 365.2422 days. During the year there are two moon-months in which eclipses can occur, and these months are called the eclipse seasons. An eclipse year containing two seasons averages 346.620 days in length.

Age observers had many hundreds, or thousands, of years in which to look, think, and record the years with tally marks.

Why have we not found such marks, evidence of calendar-keeping? It is probably because wood, bone and similar material were used; and they decay quickly in the moist British climate. Also, it is possible that the markers did not want the secret of their method known. Diodorus said ". . . the kings . . . and the supervisors of the sacred precinct [of the "spherical temple"] are called Boreades . . . and the succession of these positions is always kept in their family." On this basis one could suppose that those who had used records to devise an easily workable eclipse-prediction instrument would have taken care to protect the secret of its operation by hiding or destroying those records. Nor must we overlook the possibility that astronomical events over the years were memorized, recorded in the mind's eye. This would be critical information for preservation by the bards in their almost endless verses.

I am aware that this theorizing concerning possible use of the Aubrey holes as a computer is but speculation. My theory cannot be proved, even by the faithful 7090. The only machine that could prove it would be a time machine. But until a better theory is produced, I submit it as the most cogent solution to this problem. And it should be remarked in passing that it is most fitting that the hint which led me to this theory was provided by the man who was most responsible for their rediscovery, and their naming, forty-odd years ago—R. S. Newall.

After working out some of the many time calculations made possible by the simple moving of stones around the Aubrey circle, I tried once more to put myself back into the mind of a Stonehenge priest, or member of the family of Boreades. If I had mastered the use of that circle to track the years and the danger months, I should also want to know the days. I looked at the chart of Stonehenge and wondered how days of the month could have been marked.

The lunar month, or interval between full moons, is 29.53 days, so at first I thought of the sarsen circle, which consisted of 30 upright stones. By rotating a marker stone around them one could have followed the course of a month, approximately. But, as in the case of use of the 19-year metonic cycle to follow the swing of the moon, that 30-day moon-tracking would have soon become inaccurate. After two or three months the moving stone would have been a day off. Just as before, a longer cycle was needed. A proper day-computer should allow for the 29–30 variation.

Once more, and possibly for the final time, Stonehenge surprised me by offering a solution.

What about the 30 "Y" holes—and the 29 "Z" holes? And what about the bluestone circle, of perhaps 59 holes?

Those rings could have served as day-markers. They were about the last things built at Stonehenge, because the tracking of the days of the month is the kind of grace note which could have been left until last. To predict the year of a possible eclipse would have been the most difficult task; once the year was known, the month could have been noted by watching the direction of rise and set of the full moons, and the day foreseen by watching relative positions of sun and moon. A separate computer for counting days would have indeed been a useful luxury. Moving a stone around the bluestone circle each morning and evening would have marked an interval of 29½ days, a very fine fit to the lunar month.

An eclipse can be seen at Stonehenge only when the moon rises just before the sun sets. If the moon rises long before the sun sets the eclipse may not occur for several nights; if the moon rises *after* sunset the eclipse may have "been and gone." By carefully following the changing interval between moonrise and sunset one can predict the time of an eclipse to the hour.

The realization of this was practically forced on me during my 1964 summer vigil at Stonehenge. I knew from calculations that there would be an eclipse on a certain night. As that night approached I could not help noticing how the interval between moonrise and sunset lessened, at a steady rate of nearly an hour a day, providing a very reliable prediction tool. On the evening of the eclipse the moon rose only 15 minutes before the sun set; 6 hours later the moon began entering the earth's shadow.

It seems most probable that the Stonehengers noted and made use of that moonrise-sunset time relation to predict eclipses. Compared to the task of determining the eclipse year and month by use of Aubrey holes and rise-set directions, the foretelling of the night and the hour of the event by observation of the difference in time between moonrise and sunset would have been easy.

So much for my findings, better termed the machine's findings, at Stonehenge. I think I have demonstrated beyond reasonable doubt that the monument was deliberately, accurately, skillfully oriented to the sun and the moon. Uses of such orientation were most probably religious and agricultural.

I think I have put forward the best theory to account for the otherwise unexplained holes—the 56 Aubreys, the 59 bluestones, the 30 and 29 Y's and Z's. Such "computers" could have been used to predict those most frightening things, eclipses.

Between them, the demonstrated alignments and the theorized computer-uses account for every stone, hole, mound, archway and geometric position now marked at Stonehenge I and III. Even the strange little line of holes grouped under the designation of "A" are individually involved—the one to the north lies on the far north moon line and the other three probably measured the yearly interval of extreme moon motion during one of the cycles as it swung to the left of the heel stone.

What about Stonehenge II, the abortive double circle of bluestones? Here, unfortunately, there is too little evidence for solid theorizing. Until archaeology determines the exact number of spokes the builders intended to put in that wheel, one can only guess what its purpose might have been.

The extra pit on the axis, described in Chapter 3, destroys the pattern of 38 spokes originally proposed. My own guess is that the number of intended spokes *may* have been 37, rather than 38 or 39, and *if* it was 37 the builders may have planned to use that wheel to track the moon. On the average, the moon follows the Aubrey hole cycle, that is to say it rises over stone D in winter at intervals of 19, 19 and 18 years, not necessarily in that order, for the total of 56. If the Stonehengers wanted to count the intervals between alternate swings over D, then the number of years would be either $19+19=38$, or $19+18=37$. Thus either 37 or 38 would be a "double" period, but in practice the interval 37 occurs more frequently—on 4 out of every 5 swings on the average. The number 39 would be of no use at all in tracking the moon. If the builders did design that bluestone wheel as a moon-follower, it may be that they abandoned it so suddenly because they found that neither of the double periods, 37 or 38, followed the moon as closely as the existing Aubrey circle triple period of 56 years. Further speculation concerning this minor mystery is useless until more evidence is available.

I think there is little else in these areas that can be discovered at Stonehenge—although I must confess, as I make that flat statement, that I am filled with trepidation, and cannot forget how often the old monument has produced new astonishments.

The machine has established an extraordinary sun-moon correlation throughout the structure. Astronomy has done its best. It now rests with the prehistorians, the archaeologists, anthropologists, mythologists and other authorities to make use of these new findings to advance our understanding of the "gaunt ruin," which should no longer stand *quite* so lonely in history as it does on the great plain.

Chapter 10

THE NUMBERS GAME

Again I reported my Stonehenge findings, in articles which appeared in *Harper's Magazine* for June, 1964, and in *Nature* for June 27, 1964.

Again the response was lively, voluminous, and for the most part friendly. "Sensationelle Entdeckung eines amerikanischen Astronomen" (Sensational discoveries of an American astronomer), said the Hamburg *Echo;* "Stein zeit tempel war frühes Rechenzentrum" (Stone Age temple was an early computing device), bannered the Cologne *Rundschau.*

A Texas lady feared that by writing almost exclusively of the astronomic orientation of Stonehenge, without giving what might be called equal time to its other aspects, I might have created a misleading impression of its general purposes. My report, she wrote, "prompts visions of a future scholar's examination of the Sistine Chapel ruins. 'Ha!' he exclaims—'the place known as the Vatican was unquestionably an art school!'"

This especially appealing letter came from Washington: ". . . I was struck by the dedication of man to man actually embodied in the observations and placing of those stones . . . it is humbling to think that at Stonehenge some man—men?—existed who could not have dreamed of us as we are today but who loved us enough to wish to leave a message to us . . . the priestly-scientists of that time must have realized the uncertainty of the future and the ephemeral nature of records. Thus they put their observations into as enduring a form as they could. . . . I for one thank them for their gift."

As before, the general response contained much useful information, and several descriptions of other work, more or less similar to mine, currently being done. It never ceases to amaze me, a relative newcomer to the field, that so much ingenious theorizing and meticulous field work has been done and still is being done at Stonehenge and other megalithic monuments. I should like here to discuss a few of the more interesting such speculations now extant.

Alexander Thom, Emeritus Professor of Engineering Science at Oxford, maintains that prehistoric Britons possessed hitherto unsuspected skills in geometry. He bases his conclusion on painstaking analysis of ancient stone circles.

There are several hundred of these rings, varying in diameter from a few yards to 370 feet, scattered over England and Scotland. Called in Gaelic "tursachan" or "mourners," and in Cornwall "merry maidens," they are about 4000 years old. Some 140 of them are still in good enough condition to be studied.

Thom found that more than 100 of the "mourners" were circles, and thus uninteresting geometrically. But the rest of them were very curious. They were strange figures which at first glance looked like poorly constructed, sloppy circles, but which under close scrutiny were found to be of geometrically precise design. Most of them were composed of two disparate halves. One half was an accurate semicircle, the other was a flattened or bulging approximation of that semicircle. The flattened or bulging figures could be classified in six categories, and Thom found that he could reproduce them all, quite accurately, by simple geometric methods. All that was necessary was to lay out the "good" semicircle with a stake and a rope and then use different centers, such as the points which trisect the diameter, and different radii, such as one third of the diameter, to lay out the second, "bad" half of the figure in short arcs. One result of this asymmetric construction was that two of the six categories yielded almost circular figures whose circumferences, or peripheries, very nearly equaled exactly three times their diameters. For one group the ratio was 3.059. For the other, it was 2.957. For a true circle the circumference-diameter ratio, π, is 3.141596 . . . , a number that cannot be written exactly, which is one of the most annoying facts of mathematical life.

Were those prehistoric Britons trying to make almost-circles whose "π" equaled *exactly* 3?

Thom, speaking as an engineer, says that the differences between 3.059 and 2.957 and 3.0 are so relatively small that a modern engineer could not easily measure them in the proportions of those stone "circles," and primitive men with primitive measuring devices very probably could not have detected them. If those ancient builders *were* trying to make $\pi=3$ in their distorted circles they probably thought that they had succeeded.

Thom also maintains that many of the egg-shaped "circles" are so constructed that lines joining their various centers from which the shaping arcs were swung, and other geometrically obtainable points, form right triangles.

Finally, his analysis showed that some of the "circles" are not even modified circles. They are true ellipses. An ellipse is a fairly advanced mathematical figure. It cannot be formed by using one stake and one rope, as all of the squashed or bulging "circles" could have been. It can be formed by setting up two stakes, putting a loop of rope around them, and moving a marker around the loop. An ellipse is not an easy figure to visualize, nor to construct, but Thom thinks that the megalithic builders did both.

He concludes that the prehistoric Britons "had a good working knowledge of elementary geometry" and could measure the length of a curved line "with an accuracy better than 0.2 per cent . . . it is perhaps too much to say that they knew Pythagoras's theorem [in a right triangle, the sum of the squares of the two short sides equals the square of the hypotenuse]. . . . Nevertheless we cannot be certain. They wrote their results in stone and it is just possible that those monuments were intended to enshrine an esoteric record of their mathematical achievements."

Science writer Alexander Marshack believes that he has evidence enough to prove that prehistoric men were counting the days between full moons, and noting the phases of the moon, for thousands of years before it has commonly been thought humans were capable of such observation and deduction.

His evidence: "thousands of notational sequences found on the engraved 'artistic' bones and stones of the Ice Age and the period following, as well as on the engraved and painted rock shelters and caves of Upper Paleolithic and Mesolithic Europe."*

Archaeologists have long been puzzled by the great number of scratches and other markings on artifacts, and cave walls, dating back into the last Ice Age. It seemed obvious that the markings had some significance beyond random decoration—but what? Marshack made an analysis of "all the available published materials and artifacts of the Upper Paleolithic and a firsthand study of many of the artifacts and caves, including a 'reading' of over a thousand notational sequences with their associated art and symbol," found more 30 and

---

* Marshack's findings were summarized in a preliminary article in *Science*, November 6, 1964. They will be treated fully in a book to be published soon.

29 tallies than simple statistics seemed to account for, and came to these conclusions:

As far back as 30,000 to 35,000 years men were noting—in a wide variety of ways—the 30 or 29 days (or nights) from full moon through the three phases to full again. Since the moon's cycle is 29.5 days, the count varied between 30 and 29. Sometimes the four subunits of the cycle, the days between the phases, or weeks, were marked, and sometimes the phases themselves were indicated; in a painted notation from Azilian Spain the two crescent phases were shown by actual reproductions of crescents accurately shaped and properly oriented.

The three large caves with paintings are, in descending order of importance, Lascaux in France and Altamira and La Pileta in Spain. Lascaux is closed now because of possible damage to the paintings by tourists' breath, and Altamira is not readily accessible, but in January, 1965, my wife and I visited La Pileta.

The cave of "the small holy water font," as the Spanish may be translated, can be reached by taxi from Ronda, which itself is reached by bus from Málaga along a mountain road with 512 hairpin bends. We enjoyed the trip. The bus stopped at junctions with small trails that led off into nowhere, and peasants boarded with their produce, onions and oranges, and chickens that later ran up and down the aisle. I noticed round, flat-topped stone platforms tucked away in the valleys and immediately thought of primitive sundials or astronomical instruments. It took half an hour of language struggle to learn that they were era—threshing circles. Part of our trouble was that we mistook the word as ira, the wrath of God, with much consequent amusement to our Spanish friends.

The taxi twisted round a limestone mountain shaped like the rock of Gibraltar, and then stopped where the trail stopped on the edge of a precipice. "Shout" said a notice in four languages, and we did. The guide emerged as a small speck from a cottage in the valley and arrived panting at the cave entrance with a large key.

Inside the cave we passed by bones, broken pottery, blackened fire pits, all revealed by the glare of a hissing lamp. We saw hundreds of red and black drawings in the nooks and crannies of the limestone. There was the famous fish and the pregnant horse, and scribbling on the wall about 3 feet above floor level. My wife theorized about the latter: perhaps it was done by children? Indeed it did look like some of the bedroom walls decorated by our daughter back home. The scribbling will probably never give us much information.

But higher up on the walls of the cave it was very different. There were careful definite marks all over the place. Surprisingly enough these more numerous markings have been overlooked in the literature. Lunar counts, seasonal marks? Perhaps they are—but without a more careful study I would not care to say.

More investigation must be done before Marshack's conclusions can be thoroughly evaluated. But if he is proved correct, then one may well agree with his statement that there should be "a re-evaluation of the origins of human culture, including the origins of art, symbol, religion, rite, and astronomy, and of the intellectual skills that were available for the beginnings of agriculture." Indeed, the findings at Stonehenge have already, I believe, made such a re-evaluation necessary.

Much work has been done, particularly by Thom and Newham, to answer an intriguing question: did Britain's megalith builders use a uniform measure of length? Thom is sure that they did, in the circles which he investigated. "From careful statistical analysis of the dimensions of these circles," he has reported, "it has been definitely established that the erectors used an accurate unit of length. . . ." This unit, which he calls the "megalithic yard," he defines as 2.72 English feet. He believes that occasionally multiples or submultiples of the unit were used.

Newham has analyzed Stonehenge distances to see if there is evidence that a single unit of measure, like the "megalith yard," was used. He has concluded that both the "Roman foot" of 11.66 English inches and the "ancient Greek foot" of 12.16 inches *may* have been used in construction, but how rigorously and to what extent is not clear. For instance, the internal diameter of the sarsen circle is about 97 feet 4 inches, which is within 2 inches of 100 Roman feet, while its internal circumference is about 300 ancient Greek feet. From heel stone along the axis to the line joining station stones 91–94 is 200 ancient Greek feet. So is the direct distance from Aubrey hole 28 to 14, or 42—a quarter of the Aubrey circle. From Aubrey hole 28 to the heel stone is 400 a.G. feet. And so on. It turns out that quite a few incidental Stonehenge distances are even multiples of the old Mediterranean foot. But the most important distances seem not to conform. The diameter (288 ft.) and circumference of the Aubrey circle, the distance from sarsen circle center to heel stone (256 ft.), the sides of the station stone rectangle—these seem not to have ancient Greek, Roman, or English, or any other, foot as a common divisor. It would surprise me if they did. Remember: the first Stonehenge builders used a rectangle and a distant point, the heel stone, to create their celestial alignments. More than 300 years later (and keep in mind how difficult communication must have been in those days, even in "the family of the Boreades") the last builders duplicated those alignments with a circle and a horseshoe. Does it seem likely that designers faced with such problems of geometry and astronomy and time would have even tried to lay out distances which were exact multiples of some common measuring unit? The angles between the extreme sun and moon positions are awkward angles. Furthermore they are set by the Creator, and not rearrangeable by man. It would be next to impossible to align stones geometrically on the ground and celestially to all those positions and yet keep the distances between the stones in round numbers of a single unit of distance.

Newham and Charrière of France have both commented on the very noteworthy circumstance that the latitude of Stonehenge is practically optimum for sun-moon rectangular alignment. If the site were moved north or south by as little as 30 miles—to Oxford or to Bournemouth—the astronomic geometry would be so changed that the station stone figure would change from a rectangle to a parallelogram. And the farther north, or south, the location was moved from Stonehenge's 51°17 latitude, the more "skew" the parallelogram would become, until you reached the equator. After that, as you moved south the parallelogram would lessen its skew until you reached the southern hemisphere counterpart of Stonehenge, latitude south 51°17, the Falkland Islands and the Strait of Magellan. There of course the astronomic geometry would correspond to that of Salisbury Plain. In other words, in the northern hemisphere there is only one latitude for which, at their extreme declinations, the sun and moon azimuths are separated by 90°. Stonehenge is within a few miles of that latitude.

Since this is an interesting point, let me amplify it. Imagine that we are observing at midsummer. The solstice sun rises along the lines which are the short sides of the rectangle, 92–91 and 94–93. The summer full moon rises along the long side 93–92. The angle that separates them on the horizon is 180° minus the angle 91–92–93 which is, at Stonehenge, close to a right angle.

Now imagine that we are observing at midwinter. The solstice sun sets along the short sides of the station stone rectangle and the midwinter full moon sets along the long side 91–94. The angle between the setting sun and setting moon is the angle 92–91–94 which again is nearly a right angle.

For a perfect astronomical fit, the long sides (94–91, 93–92) should not be exactly parallel, and at Stonehenge these lines are slightly closer together at the western end, as the astronomy requires. But the mean direction of those long sides should be perpendicular to the short sides, and it is. This angle at Stonehenge is 90°.2, or ⅕ of a degree in "error."

This small difference from 90° is smaller than an average twentieth-century building error and probably could not have been noticed by the Stonehengers.

Now imagine that we make these same summer rise and winter set observations at a more northerly latitude. The summer sunrise and winter moonset would both occur farther south. Thirty miles north of Stonehenge the angle would be about 91°, and the skew of the station stone figure, no longer a rectangle, would be noticeable to any would-be designers or builders.

There is a replica of Stonehenge at Maryhill, Klickitat County, Washington, where the Stonehenge sarsens and trilithons have been duplicated realistically in tons of concrete. But Maryhill is at the wrong latitude (5° too far south), so alas, the alignments of this American version of Stonehenge do not work.

We may therefore assume that *if* the Stonehengers were aware of the effect of change in latitude on angles between sun-moon alignments, and *if* they had therefore tried to put Stonehenge at the best latitude—that is, the latitude at which those alignment angles came closest to being 90°—then they might well have thought they had succeeded.

It seems unlikely that the choice of 51°.17 as a location for Stonehenge was made by chance.

Assuming that the decision had been made to build and all European locations were equally available, then one might reasonably assume that any latitude between Scotland's John o'Groat's House and the Strait of Gilbraltar could have been chosen. That is a latitude band of 25°. Thus, the chance of coincidence in the Stonehengers' choice of latitude was about one in 25. It seems that the first builders were even more skillful than had been thought: they had laid out a geometrically-celestially elegant pattern—a major axis oriented to extreme solar horizon points, and a rectangle whose long sides were perpendicular to that axis and aligned to extreme lunar horizon points, and one of whose diagonals was aligned to another critical moon direction—and they had placed this pattern at the only latitude in the northern hemisphere which made its unique geometry possible.

The noted astronomer Gerard Vaucouleurs has taken the trouble to calculate Stonehenge shadow positions. He has determined that the heights and positions of the stones were such that at midwinter noon the shadow of the southernmost lintel of the sarsen circle would have fallen right on the center of the monument. Also, he has found that at midsummer the shadow of the sarsen lintels falls on the bluestone circle as well as can be determined from the irregular shape of the latter.

The sarsen circle and trilithons could be built to any height once the positions were fixed. Height was a free parameter for the builders. It might well be that they used this opportunity to choose height that had meaning for them, or practical significance, so perhaps these shadow conditions were created deliberately.

Dr. Gerhart Wiebe, Dean of the Boston University School of Public Communications, offered this observation: "Stonehenge makes no sense when seen from the ground. It is impressive only when seen in plan from above. But neolithic man had no airplanes from which to view his own work—therefore he may have been signalling his prowess to the powers in the sky . . . to his gods." He said that a similar example might be the colossal "serpent mound" near Peebles, Ohio, which could only be appreciated from the sky.

Among the many comments and suggestions contained in the correspondence there was raised an archaeological problem. Newall, who took part in the extensive excavations with Col. Hawley, wrote asking that I take stones G and H out of the list of now missing but presumably once present stones. Apparently the excavating team of the 1920s could not be sure that the holes were man-made, or had ever held stones. Newall expressed his present opinion that the holes were made in the chalk by natural drainage of rain water. Others have thought that tree roots were the cause.

Atkinson, however, continues to include G and H in his tally of probable stone holes—at least he has them so charted in the 1960 edition of his book *Stonehenge*. He notes that they are almost equidistant from station stone 91, and speculates on the possibility that they might have been points along the circumference of a now-vanished large circle of widely spaced sarsen stones lying just outside the Aubrey circle during the time of Stonehenge II.

I personally feel dubious about the theory that tree roots caused the disturbances in the soil, although obviously it can hardly be proved, now, that they were not. I suggest that perhaps those disturbances were impressions made by temporary placement of stones which were later moved, since the sarsen stones of Stonehenge III would have blocked the 93–H view, and G might have been moved for some other reason. If the main purpose of G was to mark midsummer sunset, as I believe, then it was more than adequately replaced by the sunset trilithon.

Retroactively, the fate of G and H will be decided by the archaeologists.

I should naturally be sorry to lose those positions, since they make possible four solstice alignments and one lunar alignment. But such a loss would not be lethal to the sun-moon alignment theory. It would only reduce the odds in favor of that theory from ten million to a little short of one million to one.

About the last direct response to my Stonehenge reports came in a spirited editorial in the British archaeological journal *Antiquity*, September, 1964.

The editorial began by condemning the authorities for allowing "strange groups of people calling themselves Druids to disport themselves at Stonehenge and practice their recently invented religious rites [at summer solstice sunrise]. We are all for strange fringe religions, if in that unreasoning way their devotees get comfort and hope, but not if their activities affect the safety of our ancient monuments." Having ticked off the wrongful permissiveness of the officials and the malpractices of the "dotty Druids Lair," *Antiquity* turned its attention to my speculation that Stonehenge had been a neolithic computer. Its attitude was skeptical, as expressed in this quotation from an article by A. P. Trotter on "Stonehenge as an Astronomical Instrument" (*Antiquity*, 1927, 42):

"It is easy to bring all sorts of theories and conjectures now that this grand and simple monument is there. We may prolong the axis to the north-east and find that it hits Copenhagen; or . . . down to the coast, passing a little to the right of the megaliths of Carnac, and out to sea to the district where the lost Atlantis may have flourished. And we may prolong controversies about it until we fill a library."†

Antiquity withheld final judgment on my computer theory, how-

† Somehow this statement reminded me of an equally true pronouncement made 300+ years ago by Inigo Jones (cited in Barclay, *The Ruined Temple* . . .): "Certainly, in the intricate and obscure study of antiquity, it is far easier to refute and contradict a false, than to set down a true and certain resolution." Actually, the axis of Stonehenge extended to the northeast passes some 100 miles north of Copenhagen, and extended to the southwest passes about 70 miles west of the westernmost tip of France.

ever, noting that although Alexander Thom had commented on it, further assessment by other astronomers and archaeologists was needed.

With that view I concur completely. More comment *is* needed. As this book has made plain (I hope), while it has been established that the placement of some positions to orient Stonehenge celestially was almost certainly deliberate, it has not been proved that there was similar astronomic significance intended in the numbering of other positions. And unless more evidence comes to light that theory never can be proved—or disproved. But more discussion could be informative and helpful.

For example, Thom's article ("Observatories in Ancient Britain," *New Scientist*, July 2, 1964) reported finding sun alignments and other evidence of observing-counting-building abilities possessed by the creators of megalithic monuments, some older than Stonehenge, and he credited those creators with "knowledge of geometry, arithmetic and astronomy. . . ." He felt that Diodorus' statement about the "spherical temple" gave "great support" to the computer theory and concluded that "independent confirmation" may some day be had from study of surveys of other large megalithic sites.

The indications are mounting that early man in Europe was more intelligent than has generally been thought—quite intelligent enough to have used numbered cycles of the moon to follow and predict eclipses. Did he—or might he have? More discussion, by all means!

I am well aware of the dangers of overspeculation concerning Stonehenge. What the *Encyclopaedia Britannica* calls "fruitless conjecture" and British archaeologist Jacquetta Hawkes calls "doubtful and indeed crazy theorizing" can indeed engender controversies which, prolonged, fill libraries.‡

There are a great many numbers and alignments at Stonehenge, and numbers and lines never cease to fascinate people. Even that most rational of the Age of Reasoners, Samuel Johnson, observed most carefully the crosslines as he walked. And one of the notorious "marvels" of modern France is the fact that Paris is so aligned that on Napoleon's birthday—August 15—the sun, as seen from the Champs Élysées, sets in the center of the Arc de Triomphe. Actually, that supposed marvel is a good example of an apparently extraordinary and speculation-worthy circumstance that is in fact not

‡ *Ecclesiastes* may still have the last and best word on this sort of activity: "Of the making of many books there is no end, and much study is a weariness of the flesh," said the Preacher.

very remarkable. Let us examine the situation closely: What are the chances of simple coincidence? First we find that the Arc is so wide that the sun sets in it for a two-week period; that reduces the odds against the event being unique to Napoleon's birthday from 365 to 26 to 1. Then we must note that the sun also sets in the Arc during a two-week period in April; the odds fall to 13 to 1. Then it must be admitted that sunrise on the same day would be equally phenomenal; reduce the odds to 6½ to 1. Then we may suppose that sunset or sunrise on the day of Napoleon's death would be equally notable; 3¼ to 1. And what if the birthday sun rose or set through some similar great arch or other Napoleonic relic? And so forth. The Napoleonic sunset clearly has no significance. I think that any good coincidentalist could find just as marvelous Napoleon magic at Stonehenge; perhaps the moon rose on the line from the center of Stonehenge passing over the battlefield of Waterloo, on the morning of the battle there. What if it did?

The numbers game is nothing but a game if played without purpose and method. But there can be good result if speculation is implemented properly.

There are doubtless many remarkable things yet to be discovered about Stonehenge and the other megalithic monuments. Any and all research into these mysteries is of course needed—IF it is carried out with as much discipline as the builders of those monuments displayed.

Chapter 11

LAST THOUGHTS

That next summer, 1964, I went again to Stonehenge. The circumstances were quite different from the 1961 visit which had begun the long investigation. Then, I was a stranger, just one more of the 300,000 tourists who come to look at the famous stones every year. Now, I felt that I was an old acquaintance, almost a friend.

A television crew had come to make a documentary film of the summer solstice events—the midsummer sunrise over the heel stone and a moon eclipse through the central trilithon. (I may say that the Aubrey circle computer would have predicted the latter event by having a stone at hole 56; previous summer solstice moon eclipses had occurred in 1945, 1926 and 1908, i.e. 19, 38, and the Aubrey circle total of 56 years before.)

The Ministry of Public Buildings and Works in London gave admirable cooperation for the venture, but the filming was not done without obstacles. Tourists were polite, but numerous, and noticeable. I remember with especial clarity a task force of four busloads of very active school children, and another unit of 40 eight-year-old girls shepherded by a minister. They were more ubiquitous, and much more visible and audible, than Hamlet's father's ghost. Gay cries, the clear songs of birds and a sonic boom almost drowned out the announcer's voice as he said his opening lines: "As we near this strange and silent place . . ."

The worst foe of all, however, was of course the weather. That English June did not seem a bad one; there *were* sunny days; but for Stonehenge observation purposes the weather was its usual wretched self—the nights and the dawns were almost without exception obscured by fog, mist or rain. Of the nine days June 19–27, only one sunrise was really clear (June 20). Not a single moonrise, sunset or moonset was clear. At Stonehenge it was hard not to rage; fierce storms would have been less infuriating than weak mists and the gentle rain. Once more I found myself admiring the Stone Age builders—and hoping, for their astronomers' sakes, that the weather then was not what it is now.

The one clear morning, June 20, was the day before midsummer and the TV men took "just-in-case-of-bad-weather-tomorrow" pictures of the sun rising almost exactly over the heel stone. The camera caught the stone, black against the lightening horizon, the sun behind it—and a crow. The bird of traditional ill omen had flapped into view at precisely the critical moment and perched on the one unique spot, the top of the heel stone; the noises that the cameraman made were practically prehistoric.

On the great day itself, June 21, sunrise was due at 4:59 British summer time (like our Daylight Saving), and the TV men were on hand long before. So were scores of curiosity-seekers, scientists, students, druids, morris dancers,* miscellaneous persons like myself, and police. There were a lot of police because of a rumor that the monument was to be honored that day by a dawn riot between groups of youths called "Mods" and "Rockers." Barbed wire had been looped around the stones, and military policemen, constables and police dogs were stationed along the sarsen circle. There was no riot, though. Four long-haired "Rockers" roared up on their motorcycles, but no "Mods" opposed them.

As the magic moment approached, the druids took over. They were allowed to go through the barbed wire entanglement to perform their sad little made-up ritual among the stones which probably were old forgotten mysteries when the real druids—the priests, teachers, healers, sacrificial murderers—came to Britain. It was an absurd, touching, pathetic performance. A harpist played in the gloom, the white-robed band saluted the heel stone and marched solemnly about waving oak leaves and incense braziers and intoning certain mumblings. And the sky grew gray.

---

* The morris dancers are the gayest, most entertaining, most attractive and most authentic group at Stonehenge during the solstice brouhaha. They are folk-dancers whose dances go back to medieval time; it is thought that they were introduced into England from Spain by John of Gaunt or from France or the land of the Flemings, now Belgium. The morris dancers themselves believe their jiggy little dance, a bit reminiscent of some Scottish reels, and some of the dances done by men in Greek *tavernas*, is descended from the old Moorish or morisco dance, the Spanish fandango. Traditionally in England the morris dance was performed by five men and a boy dressed as "Maid Marian," with two musicians. Morris dancing was at the heart of village festivities in the sixteenth century, was abolished by the Puritans, restored during the Restoration, and now flourishes internationally. "Gone, the merry morris din;/Gone, the song of Gamelyn," Keats lamented in "Robin Hood." . . . Not so.

Precisely at 4:59—right on time, to be sure—the venerable chief druid called "Arise, O sun!" "O sun" may have done just that; astronomical calculations reassured me; but there was no proof. The sky just went from gray to grayer and a cold rain began to fall on druid and cameraman alike.

As for the moon eclipse, that too occurred on schedule, 2 A.M., June 25—behind a sky not entirely opaque with fog.

After the flurry of trying to see the solstice events, I went back for a last time to Stonehenge and stood among the old stones, thinking. I thought of some of the things that many others had thought, and written, down through the centuries.

To that "perfection of the Renaissance gentleman," Sir Philip Sidney, Stonehenge was a complete and improbable mystery:

> Near Wilton sweet, huge heaps of stones are found
> But so confused that neither any eye
> Can count them, nor reason try
> What force them brought to so unlikely ground. . . .

A much more philosophic attitude toward the monument was expressed by another Elizabethan poet, Samuel Daniel. In his long didactic poem *Musophilus* he has the principal speaker thus harangue his friend Philocosmus:

> Where will you have your virtuous name safe laid?
> In gorgeous tombs, in sacred cells secure?
> Do you not see those prostrate heaps betrayed
> Your fathers' bones, and could not keep them sure?
> And will you trust deceitful stones fair laid,
> And think they will be to your honor truer? . . . .
> Poor idle honors that can ill defend
> Your memories, that cannot keep their own.
> And whereto serve that wondrous trophy now
> That on the goodly plain near Wilton stands?
> That huge dumb heap, that cannot tell us how
> Nor what, nor whence it is, nor with whose hands
> Nor for whose glory, it was set. . . .

To the eighteenth-century poet laureate–antiquarian Thomas Warton, Stonehenge was a poly-faceted puzzle. He wrote this sonnet about it:

> Thou noblest monument of Albion's isle!
> Whether by Merlin's aid from Scythia's shore,
> To Amber's fatal plain Pendragon bore,
> Huge frame of giant-hands, the mighty pile
> T'entomb his Britons slain by Hengist's guile;

Or Druid priests, sprinkled with human gore,
Taught 'mid thy massy maze their mystic lore:
Or Danish chiefs, enrich'd with savage spoil,
To Victory's idol vast, an unhewn shrine,
Rear'd the rude heap: or, in thy hallow'd round,
Repose the Kings of Brutus' genuine line;
Or here those kings in solemn state were crown'd:
Studious to trace thy wondrous origine,
We muse on many an ancient tale renown'd.

To another eighteenth-century writer, the peerless natural historian of Selborne, Gilbert White, the monument was only noteworthy as a haven for birds:

"Another very unlikely spot is made use of by daws as a place to breed in, and that is Stonehenge. These birds deposit their nests in the interstices between the upright and the impost stones of that amazing work of antiquity: which circumstance alone speaks the prodigious height of the upright stones, that they should be tall enough to secure those nests from the annoyance of shepherd-boys, who are always idling around that place."

Wordsworth of course had a great deal to say about it, in "The Prelude." Like most poets he was enchanted by the dream of the mystical, bloody druids:

. . . it is the sacrificial altar, fed
With living men—how deep the groans! The voice
Of those that crowd the giant wicker thrills
The monumental hillocks, and the pomp
Is for both worlds, the living and the dead. . . .

The stones made Sir Walter Scott, himself an active contemplator of antiquities, think of "phantom forms of antediluvian giants."

For Thomas Hardy, the temple-tomb-enigma was the symbol of destiny—mystery, love, atonement, death. Tess of the D'Urbervilles has murdered her seducer and is fleeing with Angel Clare. It is a dark night; they come to a strange "monstrous place."

" 'It is Stonehenge!' said Clare . . .

" 'The heathen temple . . . ?'

" 'Yes. Older than the centuries; older than the D'Urbervilles. . . .'"

Weary, Tess "flung herself upon an oblong slab"—the altar stone.

" 'Did they sacrifice to God here?'

" 'No.'

" 'Who to?'

" 'I believe to the sun . . .'"

She falls asleep and the night ends.

"The whole enormous landscape bore that impress of reserve, taciturnity, and hesitation which is usual just before day. The eastward pillars and their architraves stood up blackly against the light, and the great flame-shaped Sun-stone beyond them; and the stone of sacrifice midway."

The men come to take her. Tess wakes:

"'I am ready.'"†

Not all writers have found it so eerie. Logan Pearsall Smith felt right at home with, or in, Stonehenge—too much so:

"There they sit for ever around the horizon of my mind, that Stonehenge circle of elderly disapproving faces—faces of the Uncles, and Schoolmasters and the Tutors who frowned on my youth.

"In the bright center and sunlight I leap, I caper, I dance my dance; but when I look up, I see they are not deceived. For nothing ever placates them, nothing ever moves to a look of approval that ring of bleak, old, contemptuous faces."

It was the burials, "these barrows of the century-darkened dead," which impressed the World-War-I-and-after poet Siegfried Sassoon:

> Memorials of oblivion, these turfed tombs,
> Of muttering ancestries whose fires, once red,
> Now burn for me beyond mysterious glooms,
> I pass them, day by day, while daylight fills
> My sense of sight on these time-haunted hills.
>
> Could I but see those burials that began
> Whole History,—flint and bronze and iron beginnings,—
> When under the wide Wiltshire sky, crude man
> Warred with his world and augured our world-winnings!
> Could I but enter that unholpen brain,
> Cabined and comfortless and insecure,
> Ruling some settlement on Salisbury Plain
> And offering blood to blind primeval powers,—
> Dim Caliban whose doom was to endure
> Earth's ignorant nullity made strange with flowers.

I remembered, too, what others had thought its purpose and life had been; how many many theories had wreathed it, some of them centuries older than Geoffrey of Monmouth's myth that Merlin brought it "with joy" from Ireland.

† It is popularly supposed that Tess was then hung at Stonehenge, but no. She was taken to Winchester, tried, and there " 'Justice' was done, and the President of the Immortals (in Aeschylean phrase) had ended his sport with Tess."

Stonehenge: memorial to men betrayed . . . palace of Northland kings . . . temple to the Elder Gods? Buddhist shrine . . . druid altar . . . battle ring . . . queen's castle? Rendezvous of flying-saucer astronauts? Signal from earth to heaven? Burial ground . . . court of justice . . . hospital . . . market place . . . farmers' grange? city hall? schoolhouse . . . college . . . cathedral? repository of esoteric skills from lost Atlantis? sanctuary . . . place of worship of serpents, or souls . . . entrance to the world of the dead? monument to life, in the world of the living? observatory?

Some of those things it wasn't—but many it was.

How many?

For centuries it must have been an overwhelming place. Then life swirled away from it, and its uses and purposes and powers, like its dead, were forgotten. For more long centuries it stood silent, a desolation and a mystery.‡ The greatest European monument of the megalithic age, vaster than Shelley's Ozymandias and more silent, seemed destined to stand guard forever over the deep secrets of the past.

Recently, as this book has shown, a few of those secrets have been discovered. Archaeology and its sister sciences have learned some of the "who," "when" and "what" of the construction; astronomy has added information about the "why." So much, though, is still unknown.

I thought: "Remarkable though those things are which have been learned, there may be more remarkable discoveries to come." And I felt more strongly that respect for those neolithic builders which had been with me since the machine first revealed the astonishing ingenuity and accuracy of their earth-sky alignments. I thought, a little facetiously but not entirely so, "Any book about Stonehenge or any other megalithic monument should be dedicated 'To Stone Age Man —Misunderstood, Maligned, and Underestimated.'" The only reason for my not so dedicating this book is that while I *think* I do not underestimate him, nor malign him, I *know* I do not understand him. Who does?

There is still disbelief about the things that have been learned at Stonehenge. Last year a British Government official whose work has to do with ancient monuments was asked by a Boston University

‡ But not a place of terror—Stonehenge is awesome and somber, but never, not even on a moonless night, dreadful in the manner of Isaiah's Babylon, where "owls dwell . . . and satyrs dance . . . and the wild beasts . . . cry. . . ." If one were asked to describe the spirit of Stonehenge one would certainly not call it menacing or frightening; a more applicable word might be "brooding," or even "sleeping."

colleague of mine, "What do you think of Professor Hawkins' findings at Stonehenge?" He replied, "I've heard about those findings but I don't believe them. You see, the ancient Britons couldn't have been as clever as all that." Actually, between 6000 and 2000 B.C. men in various parts of the world had invented and put to use the plow, the wheel, the inclined plane, the sailboat, the lever, the arch, the processes of loom-weaving, pottery-making, copper-smelting, glassmaking and beer-brewing, to name but a few of the many evidences of "cleverness." But the old concept that all prehistoric men were clumsy hulking Neanderthalish creatures more animal than human dies hard.

The French philosopher-priest Pierre Teilhard de Chardin once declared that "organically speaking" the faculties of our remote forebears "were probably the equal of our own. By the middle of the last Ice Age, at the latest, human beings had attained to the expression of aesthetic powers calling for intelligence and sensibility developed to a point which we have not surpassed."

He had in mind of course such superb works of art as the cave paintings of Lascaux.

But I thought, as I stood there among those precise circles and those immense and delicately placed stones, that it was not only in art that those remote forebears demonstrated advanced powers of intelligence and sensibility.

We have learned much about the logical, reasoning, "scientific" abilities of the megalithic builders. The long-closed book of Stonehenge has been opened a little. Perhaps, with more exploration and investigation, and more understanding, and luck, that book may be opened further.

Ninety years ago Henry James produced one of the most evocative descriptions of Stonehenge ever written. In his day the monument was a "rather hackneyed shrine of pilgrimage," and picnic parties were given to "making libations of beer on the dreadful altar sites."

But, he wrote, "The mighty mystery of the place has not yet been stared out of countenance . . . we were left to drink deep of the harmony of its solemn isolation and its unrecorded past. It stands as lonely in history as it does on the great plain, whose many-tinted green waves, as they roll away from it, seem to symbolize the ebb of the long centuries which have left it so portentously unexplained. You may put a hundred questions to these rough-hewn giants as they bend in grim contemplation of their fallen companions; but your curiosity falls dead in the vast sunny stillness that enshrouds them, and the strange monument, with all its unspoken memories, becomes simply a heart-stirring picture in a land of pictures. It is indeed immensely picturesque. I can fancy sitting all a summer's day watching its shadows shorten and lengthen again, and drawing a delicious contrast between the world's duration and the feeble span of individual experience. There is something in Stonehenge almost reassuring; and if you are disposed to feel that life is rather a superficial matter, and that we soon get to the bottom of things, the immemorial gray pillars may serve to remind you of the enormous background of Time."

One could not better that Jamesean description, of course. But from the vantage point of nearly a century of theorizing, scientific investigation and machine testing, one might disagree a little with the Jamesean conclusions. Curiosity no longer falls entirely dead; some of the hundred questions have been answered, and more may follow.

> Ill did those mighty men to trust thee with their story;
> That has forgot their names who reared thee for their glory. . . .

Thus did Drayton sum up the matter. But as time passes, his conclusion too may be found in error. The names of the mighty men who built Stonehenge may indeed be forever forgotten, but their story is still being read, and interpreted, and, more and more, remembered in the stones today.

Appendix A

STONEHENGE DECODED

by Gerald S. Hawkins

Much excellent archaeological work has been done on Stonehenge, particularly by R. J. C. Atkinson[1] and others[2]. It has been established that there was building activity from approximately 2000 B.C. until 1500 B.C. At the beginning of this period the 56 Aubrey holes (Fig. 1) were dug at equal spacings around a circle with errors of less than 0.5°. At the final phase the giant trilithon archways were in position, surrounded by the sarsen circle. The heel stone and four station stones (91, 92, 93, and 94) were set in position some time before the building of the central monument.

Little astronomical work has been done on the ancient structure. For years it has been popularly thought that its major axis, the avenue, points to the midsummer sunrise, and in 1901 Sir Norman Lockyer[3] tried to estimate the date of construction by applying astronomical calculation to that assumption. He was justifiably criticized for this procedure[1,2] because we have no record of what the ancients took to be the instant of sunrise. Was it the first gleam or the moment when the whole disk stands on the horizon? We do not know. Since 1901 there has been no major astronomical investigation. This article presents some astronomical findings which I have recently made.

Assuming a construction date of 1500 B.C. and using an IBM 7090 electronic computer, significant horizon positions for rising and setting of Sun, Moon, stars and planets were determined. Positions of the Sun were for midsummer, northernmost declination, and midwinter, southernmost (approximate declinations noted on plan; Fig. 1). Since nodal regression caused the maximum declination of the full Moon to vary between 29.0° and 18.7° north and south in an approximately 9-year period, the four positions of the Moon were examined. Rising and setting was taken as the point where the disk stands tangent on the horizon. The apparent altitude of the horizon was taken as 0.6° and atmospheric refraction as 0.47°. The parallax of the Sun and Moon was taken as 0.0025° and 0.9508°, respectively.

This article originally appeared in *Nature*, October 26, 1963. It is reprinted here by permission of *Nature*, Macmillan (Journals) Limited, London.

Then positions of all stones, holes and midpoints were measured. For this I used two sketches. The first[1] is drawn to a scale of about 40 ft. to the inch. The second, kindly supplied by Mr. B. V. Field of the Ministry of

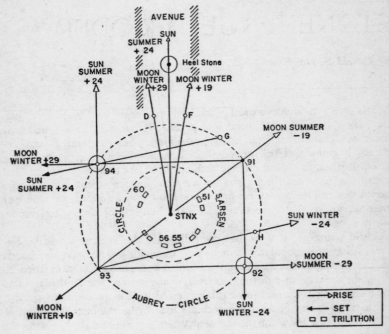

Fig. 1. Schematic plan of Stonehenge

Public Building and Works, is scaled 20 ft. to the inch. Random azimuthal differences of approximately 0.2° were found between the two sketches, and part of these differences could have been caused by my probable measurement error. Since the second sketch is of larger scale and more recent date, its values were assumed. Holes F, G, H were transferred from the first sketch. Holes were measured from the centre, missing stones were measured by estimate from adjacent stones. The original spacing of the great trilithon was taken as 30 in. Identification in the accompanying plan is according to accepted convention. STNX is the intersection of station stone diagonals. The heel stone, 92 and 94 are circled to indicate mounds.

The reference azimuth is the line from the heel stone through the nearest sarsen archway and STNX. From Lockyer's survey this azimuth is 51.23° east of north. By ciné film measure of a sunrise, I obtained a value differing by only 0.15°; in this work I have used Lockyer's figure.

The machine programme called for the positions of stones, stone holes, etc., in selected pairs, and the azimuths and horizon declinations were computed. These alignments were then compared with the positions of the celestial bodies, and the errors of alignment computed.

Stars and planets yielded no detectable correlation. For the Sun and Moon, the results of the machine testing were remarkable and are shown in Tables 1 and 2. To a mean accuracy of 1° the Sun yielded 10 correlations; to a mean accuracy of 1.5° the Moon gave 14. The correlations for the station stones are shown in Fig. 1.

The mean horizontal error (Tables 1 and 2) in the station stones is 80 in., and in the sarsens 20 in.; but these are not necessarily building errors. For example, the heel stone is now leaning at an angle of 25°. In 1500 B.C. its top lay some 21 in. below the lower limb of the rising midsummer Sun; but if the stone were set upright this vertical error would disappear. The Moon was difficult to observe because of the variation from year to year. If the midwinter full Moon was obscured by cloud, for ex-

TABLE 1
STATION STONE DIRECTIONS

Stone	Seen from	Azmth. (deg. N.)	Object	Decl. (deg.)	Error Alt. (deg.)	Error Hrzt. (in.)	Error Vrt. (in.)
94	93	51.5	Midsummer sunrise	+23.9	+0.7	−24	+15
Heel	STNX	51.3	Midsummer sunrise	+23.9	+0.5	−45	+28
94	G	309.4	Midsummer sunset	+23.9	+0.1	+10	+6
H	93	128.2	Midwinter sunrise	−23.9	−1.7	+159	−100
92	91	229.1	Midwinter sunset	−23.9	+0.0	+1	+1
A	STNX	43.7	Midwinter moonrise	+29.0	+1.0	−90	−51
D	STNX	43.7	Midwinter moonrise	+29.0	+1.0	−58	+33
94	91	319.6	Midwinter moonset	+29.0	−0.8	−76	−40
F	STNX	61.5	Midwinter moonrise	+18.7	+0.3	−12	+8
93	91	297.4	Midwinter moonset	+18.7	+1.1	+90	+65
92	93	140.7	Midsummer moonrise	−29.0	−1.3	+145	−73
Not marked			Midsummer moonset	−29.0			
91	93	117.4	Midsummer moonrise	−18.7	−3.8	+315	−223
Not marked			Midsummer moonset	−18.7			

ample, when the declination was $+29°$, then the measured value in the preceding or following year would be 0.5° smaller. Thus when the declination is positive the error in altitude should be positive and vice versa. As can be seen from Tables 1 and 2, this is true for all alignments except 94, which incidentally is an un-excavated hole, and there may be uncertainty in the position.

From Bernouilli's theorem, the probability that these ten positions are marked by chance alignment in the two structures seems less than one in a million.

I believe that this is the first time such correlations have been determined in detail—possibly because the magnitude of the task has deterred workers without access to machine computers.

A full description of this investigation is in preparation and will be published elsewhere. Meanwhile it is of interest to summarize some of the more important deductions.

TABLE 2
TRILITHON AND SARSEN DIRECTIONS

Stone	Seen from	Azmth. (deg.)	Object	Decl. (deg.)	Error Alt. (deg.)	Error Hrzt. (in.)	Error Vrt. (in.)
Heel	30–1	51.2	Midsummer sunrise	+23.9	+0.5	−34	+21
23–24	59–60	304.7	Midsummer sunset	+23.9	+3.2	+26	Missing
6–7	51–52	131.6	Midwinter sunrise	−23.9	+0.4	−3	
16–15	55–56	228.9	Midwinter sunset	−23.9	−1.4	−11	Missing
55–56	STNX	226.7	Midwinter sunset	−23.9	+1.4	+14	
A	30–1	41.6	Midwinter moonrise	+29.0	−0.1	+9	−5
D	30–1	39.9	Midwinter moonrise	+29.0	−1.0	+42	−22
21–22	57–58	315.2	Midwinter moonset	+29.0	+1.7	+16	
F	1–2	60.4	Midwinter moonrise	+18.7	−0.5	+13	−9
20–21	57–58	292.0	Midwinter moonset	+18.7	+5.1	+35	Missing
9–10	53–54	139.4	Midsummer moonrise	−29.0	−2.0	+22	
Not marked			Midsummer moonset	−29.0			
8–9	53–54	120.6	Midsummer moonrise	−18.7	−1.5	+13	
Not marked			Midsummer moonset	−18.7			

(1) Observations by Stonehenge people were not made from the mounds; the rising or setting over a mound was viewed from the distant stone.

(2) Midsummer sunrise and midwinter sunset are not diametrically opposite; the angle is about 178°, depending on the altitude of the horizon. The avenue axis is a 'line of best fit', approximately perpendicular to the bisector of that angle. This accounts for the off-set of the heel stone. Lockyer's statement that the avenue marked the first gleam of sunrise is substantially correct for 1500 B.C.

(3) When the sarsen ring was built, most of the previous sighting lines of the station stones were preserved. Lines 94–91, 92–93 have a 2-ft. clearance, for example. However, 91–93 and H-93 were blocked, but these were the worst sighting lines of the station stones, and were replaced by the trilithons.

(4) The Aubrey holes do not mark specific risings or settings. This circle probably provided an accurate protractor for the initial measurement of azimuth, the raised bank providing an artificial horizon.

(5) Although the sarsen circle and trilithons are symmetrical, there is no astronomical symmetry about the chosen axis. Thus the missing sarsen stones would not mark the related rising and setting very well. Perhaps these stones were never erected.

To determine the anthropological reason for Stonehenge is impossible, and one can only speculate. The monument could certainly form a reliable calendar for predicting the seasons. It could also signal the danger periods for an eclipse of Sun or Moon. It could have formed a dramatic backdrop for watching the interchange between the Sun, which dominated the warmth of summer, and the Moon, which dominated the cold of winter.

This work was made possible by the donation of approximately one minute of time on the Smithsonian-Harvard electronic computer.

References

[1] Atkinson, R. J. C., *Stonehenge* (Hamish Hamilton, London, 1956).
[2] Stone, E. H., *The Stones of Stonehenge* (Robert Scott, London, 1924).
[3] Lockyer, N., and Penrose, F. C., *Proc. Roy. Soc.*, **69,** 137 (1901).

Appendix B

STONEHENGE:
A NEOLITHIC COMPUTER

by Gerald S. Hawkins

Diodorus in his *History of the Ancient World*,[1] written about 50 B.C., said of prehistoric Britain: "The Moon as viewed from this island appears to be but a little distance from the Earth and to have on it prominences like those of the Earth, which are visible to the eye. The account is also given that the god [Moon?] visits the island every 19 years, the period in which the return of the stars to the same place in the heavens is accomplished. . . . There is also on the island both a magnificent sacred precinct of Apollo [Sun] and a notable temple . . . and the supervisors are called Boreadae, and succession to these positions is always kept in their family."

I am indebted to the British archaeologist R. S. Newall for directing my attention to this classic work. The statement of Diodorus is second-hand and has sometimes been dismissed as a myth, but there is a possibility that it refers to Stonehenge.

The Moon rises farthest to the north when it appears over stone D as seen from the centre of Stonehenge,[2] similar to the rising of the midsummer Sun over the heel stone. In a period of 18.61 years the extreme moonrise will shift from D to the heel stone to F and then return to D. The extreme moonrise thus swings from side to side in the avenue because of the regression of the nodes. When we consider a particular moonrise, such as the nearest full moon to the winter solstice, which we will call 'midwinter moonrise,' then the cycle takes either 19 or 18 years.

The position of the Moon has been computed using first-order terms[3] from 2001 to 1000 B.C. and the azimuth of moonrise has been determined for each winter solstice during this period. A sample of the results from 1600 to 1400 B.C. is shown in Fig. 1. Mrs. S. Rosenthal assisted with the programming of the I.B.M. 7094, and I thank the Smithsonian Astrophysical Observatory for the donation of 40 sec of machine time for this problem.

"Stonehenge: A Neolithic Computer" originally appeared in *Nature*, June 27, 1964. Reprinted here by permission of *Nature*, Macmillan (Journals) Limited, London.

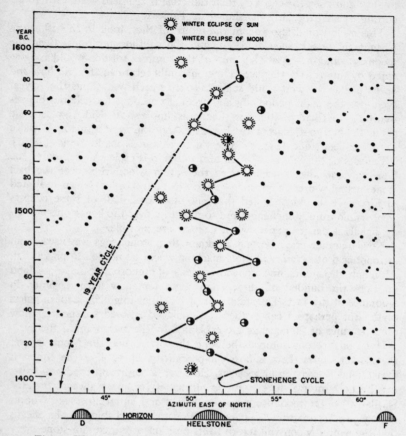

Fig. 1. The azimuth of winter moonrise from 1600 to 1400 B.C.

With midwinter moonrise the cycle is primarily one of 19 years with 38 per cent irregularity. For example, the Moon rises over F in 1671, 1652, 1634, 1615, and 1596 B.C. The intervals are 19, 18, 19 and 19 years respectively. Actually, from 2001 to 1000 B.C. the winter Moon is over F 52 times, and there are 32 intervals of 19 years and 20 of 18 as shown in Table 1. Similarly the cycle is primarily one of 19 years for moonrise over D at the winter solstice (Table 1).

The winter Moon rises over the heel stone with twice this frequency. For example, in 1694, 1685, 1676 and 1666 B.C. the intervals are 9, 9 and 10 years. Over the period 2001 to 1000 B.C. the '10' irregularity occurs with a frequency of 33 per cent. However, if we consider second intervals, 1694 to 1676 and 1685 to 1666 B.C., then the cycle is again 19 years with 18 occurring as an irregularity as shown in Table 1.

This cycle would also govern the return of the Moon to the other important alignments such as 94–91, and the trilithon positions. Even the moonrise along 92–93 at the time of the summer solstice would be governed by this 19, 19, 18 cycle. The Sun would return to the trilithon and heel stone at the winter and summer solstice each year. Thus the 19-year cycle was the main periodicity and seems to account for celestial objects returning to their positions as Diodorus implies. A rigid 19-year cycle gradually becomes inaccurate, however, and the winter moon deviates from the heel stone (Fig. 1) unless a correction is made every 56 years.

Eclipses of the Sun and Moon also follow this cycle. An eclipse of the Sun or Moon always occurs when the winter Moon rises over the heel stone; actual winter eclipses[4] from 1600 to 1400 B.C. have been indicated in Fig. 1. It should be noted that not more than half of these eclipses were visible from Stonehenge, and so moonrise over the heel stone primarily signals a danger period when eclipses are possible.[2]

Now I cannot prove beyond all doubt that Stonehenge was used as an astronomical observatory. A time machine would be needed to prove that. Although the stones line up with dozens of important Sun and Moon positions the builders of Stonehenge might somehow have remained in ignorance of this fact. The statement of Diodorus might be a meaningless myth. But perhaps I can reduce the doubt to a shred by showing how other features of Stonehenge are explained by the astronomical theory.

If we take second intervals between the years when the Moon is over the marker stones there is no clear periodicity; in Table 1 the Moon is over D and F every 37 or 38 years. However, a surprising condition exists for the next interval in extreme azimuths—it is almost always 56 years! Similarly, winter moonrise over the heel stone and eclipses also occur exactly 56 years apart on 84 per cent of all occasions (Table 1). This means that the winter Moon will return to its position over a certain stone every 56 years, and there are many such cycles which will become due in the span of a human lifetime. For example, during 20 years of observation the Moon would take up the ten positions which I have noted[2] in both the sarsen circle and station stones. Each of these occurrences would have been a part of a sustained 56-year cycle and therefore could have been predicted by a person with knowledge of the cycle—knowledge "kept in their family" as Diodorus says.

The number 56 is of great significance for Stonehenge because it is the numbers of Aubrey holes set around the outer circle. Viewed from the centre these holes are placed at equal spacings of azimuth around the horizon and, therefore, they cannot mark the Sun, Moon or any celestial object. This is confirmed by the archaeological evidence; the holes have held fires and cremations of bodies, but have never held stones. Now, if

TABLE 1

INTERVAL IN YEARS BETWEEN WINTER MOONRISE OVER STONES
D, F AND THE HEEL STONE

Interval (years)	Frequency of interval (stone F)	Frequency of interval (stone D)	Frequency of interval (heel stone)
8	0	0	2
9	0	0	70 (65%)
10	0	0	35
18	20	20	40
19	32 (62%)	33 (62%)	66 (62%)
37	39 (77%)	40 (77%)	80 (77%)
38	12	12	24
54	0	0	1
55	8	8	15
56	42 (84%)	43 (85%)	86 (84%)

the Stonehenge people desired to divide up the circle why did they not make 64 holes simply by bisecting segments of the circle—32, 16, 8, 4 and 2? I believe that the Aubrey holes provided a system for counting the years, one hole for each year, to aid in predicting the movement of the Moon. Perhaps cremations were performed in a particular Aubrey hole during the course of the year, or perhaps the hole was marked by a movable stone.

Stonehenge can be used as a digital computing machine. One mode of operating this Stone Age monument as a computer is as follows:

Take three white stones, a, b, c, and set them at Aubrey holes number 56, 38 and 19 as shown in Fig. 2.

Take three black stones, x, y, z, and set them at holes 47, 28 and 10.

Shift each stone one place around the circle every year, say at the winter or the summer solstice.

This simple operation will predict accurately every important lunar event for hundreds of years. For example, to the question: "When does the full Moon rise over the heel stone at the winter solstice?", the answer is: "When any stone is at hole 56." (Hole 56 is a logical marker because it lines up with the heel stone as viewed from the centre.) In Table 2, I have given the critical years as predicted by the Stonehenge computer for the period 1610 to 1450 B.C. with the stones set so that 'a' was at hole 56 in 1610. This period was chosen because 1600 B.C. is the earliest year for which eclipses have been computed.[4] Table 2 shows the remarkable accuracy of the Stonehenge computer. The correct year was predicted on 14 occasions out of 18 and the maximum error was only one digit. It also gave the years when the nearest full Moon to mid-summer set through the great trilithon (55–56). Incidentally, a stone was at hole 28 at this time, lining up with the great trilithon.

The stones at hole 56 predict the year when an eclipse of the Sun or Moon will occur within 15 days of midwinter—the month of the winter Moon. It will also predict eclipses for the summer Moon. In 1500 B.C. the winter solstice occurred on January 6, Julian calendar, and so the 30 days between December 22, 1501, and January 21, 1500, were the period of the winter Moon. Similarly, the summer Moon and other seasons in 1500 B.C. occurred 15 days late by our present Gregorian calendar. Table 2 gives actual eclipse data, showing how Stonehenge scored 100 per cent success in predicting winter and/or summer eclipses. When more than one eclipse occurred, only one is listed in Table 2.

TABLE 2

WINTER MOONRISE OVER THE HEEL STONE AND ECLIPSES AT THE SUMMER AND WINTER SOLSTICES

Stonehenge cycle Year B.C.	Moon over heel B.C.	Lunar eclipses	Solar eclipses
1610	1610	No data available[4]	
1601	1601	No data available[4]	
1592	1591	Jul. 14, '92	Dec. 24, '92
1582	1583	Dec. 30, '83	—
1573	1573	—	Jan. 4, '73
1564	1564	Jan. 10, '64	—
1554	1554	—	Jan. 4, '54
1545	1545	Jan. 10, '45	—
1536	1536	—	Jan. 14, '36
1526	1527	Jul. 16, '27	Jun. 21, '56
1517	1517	Dec. 31, '18	—
1508	1508	—	Jan. 5, '08
1498	1498	Dec. 31, '99	—
1489	1489	—	Jan. 6, '89
1480	1480	Jan. 10, '80	Jun. 21, '80
1470	1471	Dec. 22, '71	Jul. 12, '71
1461	1461	—	Jun. 21, '61
1452	1452	Jan. 1, '52	Jul. 12, '52

To summarize the mode of operation for the reader, the six movable stones give intervals of 9, 9, 10, 9, 9, 10, . . . years after 1610 B.C. The a, b, c stones give intervals of 18, 19, 19, . . . years. The Stonehenge cycle keeps in step with the Moon because it gives an average period of 18.67 years and the regression of the nodes of the Moon's orbit is close, 18.61 years. It keeps in step with eclipses because the metonic cycle of 19 years and the saros of 18 years are both eclipse cycles. The metonic cycle has not been previously recognized as an eclipse cycle, probably because it runs for only 57 years or so. It is, however, a remarkable cycle because eclipses repeat on the same calendar date. The lunar eclipse of December 19, 1964, for example, follows the lunar eclipse of December 19, 1945.

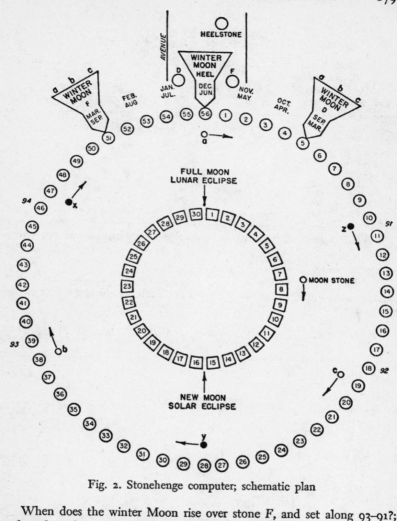

Fig. 2. Stonehenge computer; schematic plan

When does the winter Moon rise over stone *F*, and set along 93–91?; when does the summer Moon rise over 91 as seen from 93?; when does the equinox[5] Moon rise and set along 94–C, and when do eclipses occur at the equinoxes? Answer: When a white stone is at hole 51. A comparison of the Stonehenge prediction and the actual dates is given in Table 3. Again the accuracy is very satisfactory.

When does the winter Moon rise over stone *D*, and set along 94–91?; when does the summer Moon rise over mound 92 as seen from 93?; when does the equinox Moon rise and set along 94–C, and when do eclipses oc-

TABLE 3
WINTER MOONRISE OVER STONE F, AND ECLIPSES OF THE
HARVEST AND SPRING MOON

Stonehenge cycle Year B.C.	Moon over F B.C.	Lunar eclipses	Solar eclipses
1597	1596	Apr. 13, Oct. 6, '97	Mar. 18, '96
1578	1578	Apr. 13, Oct. 7, '78	—
1559	1559	—	Mar. 29, Sep. 22, '59
1541	1540	—	Apr. 9, Oct. 2, '41
1522	1522	—	Apr. 9, Oct. 3, '22
1503	1503	Mar. 25, '03	Apr. 9, Oct. 3, '03
1485	1485	Apr. 4, Sep. 28, '85	Apr. 19, Oct. 13, '85
1466	1466	Apr. 5, Sep. 29, '66	—
1447	1447	—	Mar. 20, '47

cur at the equinoxes? The answer to all these questions is: When a white stone is at hole 5. A sample run (Table 4) shows the accuracy of the stone machine.

Needless to say, Tables 2, 3 and 4 also predict the appearances of the moonrise and moonset in the trilithon and archways of the sarsen circle,

TABLE 4
WINTER MOONRISE OVER STONE D, AND ECLIPSES OF HARVEST
AND SPRING MOON

Stonehenge cycle Year B.C.	Moon over D B.C.	Lunar eclipses	Solar eclipses
1605	1606	No data available[4]	
1587	1587		Apr. 7, Oct. 1, '87
1568	1568	Mar. 23, '68	Apr. 7, '68
1549	1550	Mar. 23, '49	—
1531	1531	Apr. 3, Sep. 28, '31	
1512	1512	—	Mar. 20, Oct. 12, '12
1493	1494	—	Mar. 19, Sep. 24, '94
1475	1475	—	Mar. 19, '93, Sep. 24, '94
1456	1457	—	Mar. 30, Sep. 23, '56

because this later construction repeats the 10 lunar-solar alignments of the station stones.

In what years will eclipses occur between the solstice and equinox? In terms of our calendar, take the months of April and October as an example. When any stone is at holes 3 or 4, eclipses occur during these months. The sector between 51 and 5 has been marked appropriately in the diagram so that it predicts the eclipse seasons according to our present-day calendar.

One remaining requirement was to be able to determine which full Moon was nearest to the solstice or equinox. The average time between one full Moon and the next is 29.53 days and the Stonehengers would need to count that interval. A movable stone in the 30 archways of the sarsen circle would be sufficient. If it were moved by one position each day, full Moon could be expected when the stone was at a particular archway, such as 30–1. The stone would require resetting by ± 1 position every two or three months to stay in time with the somewhat irregular Moon. As the solstice or equinox approached (shown by solar observations), the Stonehenger could decide which full Moon was going to be the critical one. The sarsen circle could also have been a vernier for predicting the exact day of an eclipse. A lunar eclipse occurs when the Moon stone is in archway 30–1; a solar eclipse when the Moon stone is in 15–16.

A complete analysis shows that the stone computer is accurate for about three centuries, and then the Moon phenomena will begin to occur one year early. This would be noticed by the Stonehengers and could have been corrected simply by advancing the six stones by one space. The process is known today as resetting or recycling, and is used by all modern computers and logic circuits. A simple rule to add to the operating instructions would be to advance all six stones by one hole when the Moon phenomena are a year earlier than the prediction of a particular stone, say stone a. This is not a critical adjustment. If the error was not noticed with stone a, because of clouds for example, the error could still be corrected with the following stones, x, b, y, etc. The adjustment becomes due once every 300 years or so, in 2001, 1778, and 1443 B.C., for example.

Precession does not affect the accuracy, and the change of obliquity of the ecliptic and Moon's orbit also have very little effect. In 1964, for example, stone a is at 56. The full Moon rises over the heel stone on December 19, will be eclipsed at 2.35 a.m., and will set along 94–G. The next winter eclipse is also visible at Stonehenge, and is marked by stone x, 9 years later on December 10, 1973. The Stonehenge computer will function until well beyond A.D. 2100, when it will require resetting by one hole. It will then function for at least another 300 years before further resetting is required.

References

[1] *Diodorus of Sicily*, Book II, **47** (Harvard Univ. Press, Cambr., 1935).

[2] Hawkins, G. S., *Nature* **200**, 306 (1963).

[3] *Explanatory Supplement to the Astronomical Ephemeris* (H.M.S.O., London, 1961).

[4] Van den Bergh, G., *Eclipses* —1600 to —1207 (Tjeenk, Willink and Zoon, Holland, 1954).

[5] Newham, C. A., *The Enigma of Stonehenge* (private publication, 1964).

CALLANISH, A SCOTTISH STONEHENGE

A group of standing stones was used by Stone Age man to mark the seasons and perhaps to predict eclipse seasons

Gerald S. Hawkins

The stones and archways at Stonehenge point to the sun and moon as they rise and set during the year (1). Between winter and summer the sun rises further to the north every day, and the extreme position on midsummer's day is marked by the heel stone. The heel stone was placed with an accuracy of better than 0.2°, a remarkable precision for the period (2000–1500 B.C.). Between summer and winter the sun rises further to the south every day, and its extreme southern position on midwinter's day is marked by archways in the structure. The rising and setting of the sun at the equinoxes are also marked. Thus, altogether six solar directions are marked.

In a similar way the moon rises at a different point on the horizon every night, but the moon swings from its northern extreme to its southern extreme much faster than the sun does. The moon takes 2 weeks to complete its swing, whereas the sun takes 6 months. For the moon there is a further complication—the slow wobble of its orbit. Without this wobble the full moon nearest midwinter's day would rise over the heel stone every year, and the moon would be furthest north on the horizon at this time. Because of the wobble, the midwinter full moon swings first to the left and then to the right of the heel stone through an angle of about 20°. The moon requires 18.61 years to complete one cycle, and it requires almost exactly 56 years to complete three cycles. The swing of the moon provides 12 extreme positions of the full moon on the horizon that could have been marked by the Stone Age astronomers, in summer and winter,

"Callanish, a Scottish Stonehenge" originally appeared in *Science*, Vol. 147, No. 3654, January 8, 1965, pp. 127–130. Reprinted here by permission of *Science*.

and at the equinoxes—two extreme positions for each of the six extreme positions of the sun. Figure 1 shows these directions for the latitude of Stonehenge, 51°N. (The equinox alignments are unpublished.)

When the full moon rises opposite the setting sun, an eclipse of the moon is possible. An eclipse of the sun may occur 15 days later, when the moon has moved around its orbit to line up with the sun. The periods in which eclipses are possible are known as "eclipse seasons." Their occurrence in the calendar is controlled by the 18.61-year cyclic precession of the moon's orbit, and an eclipse year of 346.620 days contains two eclipse seasons. After 56 years the sequence of eclipse seasons returns to within 3 or 4 days of the starting point in the Gregorian calendar. This fact is confirmed by the commensurate length of 56 tropical years and 59 eclipse years. This is the eclipse cycle which synchronizes most accurately with the tropical year, with a period of less than 90 years.

I have suggested (2) that the 56 Aubrey holes at Stonehenge were used to predict the eclipse seasons. These are set at equal spacings around a perfect circle. Each hole was dug into the chalk to a depth of about 1½ meters and then refilled with white chalk rubble. Cremated human remains were later placed in the holes, a finding which lends support to the archaeological opinion that the holes were ritual pits. By moving marker stones around the circle, changing the position by one Aubrey hole each year, the Stonehengers could predict the particular year in which there would be danger of, say, eclipse of the winter moon. By means of the 30 archways, the Stonehengers could predict the actual day of an eclipse. The archways were set in a perfect circle within the circle of Aubrey holes, and I have suggested that each gap represented a day of the lunar

TABLE 1
ASTRONOMICAL ALIGNMENTS AT CALLANISH

Object	Stone	Point viewed from	Azimuth (deg N)	Declination (deg)	Altitude of the horizon (deg)	Error (deg)
Rising midsummer sun	34	29	41.8	+23.9	0.8	+0.2
Setting midsummer sun	20	9	316.2	+23.9	.3	+1.4
Rising sun at equinox	20	23	91.5	+0.0	.8	+0.3
Rising midsummer moon	35	29	163.9	−29.0	.5	+0.1
Setting midsummer moon	10	18	190.1	−29.0	1.3	+0.1
Setting midsummer moon	1	7	191.4	−29.0	1.3	+0.0
Rising midwinter moon	30	35	26.6	+29.0	1.7	−1.4
Rising midwinter moon	33	35	56.0	+18.7	1.0	+0.0
Rising midwinter moon	34	9	32.5	+29.0	1.3	+0.6
Setting moon at equinox	30	33	259.1	−5.2	1.0	−1.0
Midsummer moon at transit	24	28	182.0	−29.0	0.6	1.25

month. By moving a marker stone from one archway to the next each day, a person could follow the phases of the moon and predict the danger of a lunar eclipse, which takes place only at full moon, and a solar eclipse, which occurs at the "new" phase. By observing whether or not the moon rose before the sun set, a Stonehenger could estimate the local time of an eclipse to within an hour. Thus, Stonehenge may well have been a device of such precision and complexity of design as to indicate a level of intellect far surpassing that which we have hitherto been willing to ascribe to Stone Age man.

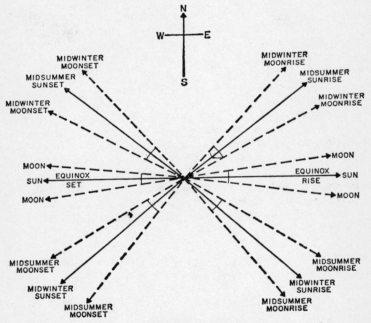

Fig. 1. The azimuthal direction of the rising and setting of the sun and moon at solstice and equinox for the latitude of Stonehenge.

Stonehenge is a very special monument with no exact counterpart anywhere in the known world. One might expect, however, to find that other stone circles built around 2000 B.C. had a similar astronomical function. As the British archaeologist R. S. Newall said, "I don't fancy it [the proposed astronomical function for Stonehenge] will be accepted by archaeologists until other sites that could be used in a similar way are found in Britain or on the Continent."

Callanish

Few of the plans of the several hundred megalithic monuments and stone circles in Great Britain have been published, but Somerville has published one (3), that of Callanish (Fig. 2). Callanish is a group of large standing stones situated on Lewis, the northernmost island of the Outer Hebrides, a rather desolate spot some 130 kilometers north of Barra. Callanish consists of a ring of 13 stones with a central great stone, an avenue, and other deliberately set rows. Somerville suggested that the avenue was aligned to point to the star Capella at its rising, and that the four stones to the east of the avenue pointed to the rising Pleiades. But a star, as viewed at sea level under even the very best conditions, is less bright by at least six magnitudes than it is when viewed higher in the sky, and Capella at its rising would be faint and inconspicuous. The Pleiades would be invisible to the naked eye. Somerville also suspected one moon alignment, however, and so Callanish becomes a prime candidate for study in the search for megalithic sites that could have been used in ways similar to those proposed for Stonehenge.

The position of all the stones of Callanish was read by Julie Cole, using a rectangular grid, and the azimuths of the lines between any two stones were computed. The azimuth for stone 20 as seen from stone 23 was taken to be $91°48$, an error of $0°58$ in Somerville's published plan, as reported by Thom (4), being taken into account. The altitude of the horizon was calculated from contours on the 1-inch (2.5-cm) Ordnance survey map. Allowance was made for atmospheric refraction and parallax in calculating the declination of an object on the horizon.

At Callanish, ten alignments with the sun and moon at their extreme positions on the horizon were found. Furthermore, as inspection of Fig. 2 shows, these alignments are the most important ones in the structure. The error in the setting of the stones is given in column 7 of Table 1. It is expressed as height above the horizon, at sunrise or sunset (or moonrise or moonset), of the lower limb of the sun (or moon) as seen along the line of stones. Errors were found to be minimal when a definition of sunrise and sunset as the time when the lower limb is tangential to the horizon was assumed. This definition of sunrise and sunset seems to have been used by the Stonehengers, particularly with the heel stone, as well as by the people of Callanish.

The latitude of Callanish is of some interest. It is near the Arctic circle for the moon, the latitude where the moon at its extreme declination remains hidden just below the southern horizon. Callanish is $1°3$ south of this critical latitude, and there the full moon at midsummer stands about $1°$ above the southern horizon every 18 or 19 years. The row of stones from 24 to 28 points to the rising, transit, and setting of the moon along its path at these times, when it appears to come closest to the horizon. Midsummer moonset is over Mount Clisham, the highest peak on Harris, and the avenue points to this mountain. Perhaps this alignment of the moon with the mountain was significant for the Callanish people.

The eastern triangle of stones, with apexes at stones 30, 33, and 35, is interesting. As viewed from stone 35, the swing of the midwinter moon from declination +18°7 to +29°0 is marked by the row of stones 30 to 33. On the average, the midwinter moon stays 3 years in each of the three gaps in this row.

Stone 35, in alignment with a second stone, marks three different lunar directions (see Table 1). Most of the stones listed in Table 1 mark at least two lunar or solar directions. This gives added weight to the theory that the astronomical alignments were intentional.

The error in the setting of the alignments is about 0°5 in altitude. That is to say, the lower limb of the sun or moon was about one-half degree above the point on the horizon to which the line of stones was directed. This is considerably better than the accuracy at Stonehenge, but the greater accuracy is largely attributable to the high latitude. The six directions of the rising or setting sun and the 12 directions of the rising or setting moon are shown for Callanish in Fig. 3. The directions are different from those at Stonehenge (Fig. 1) because of the difference in latitude. The sun (or moon) when rising and setting follows a more slanting path as it crosses the horizon at Callanish than it does at Stonehenge. The path of the midsummer moon, computed for 1500 B.C., is shown in Fig. 4. At Callanish a large change in azimuthal bearing of the sun produces a small change in altitude above the horizon. Thus, the error in azimuthal bearing is about the same as that at Stonehenge. At least some of the errors given in Table 1 arise from errors in the available chart of the structure, from which calculations were made, and from uncertainties concerning the elevation of the horizon. Before a detailed discussion of errors is undertaken Callanish must be resurveyed and measurements must be made of the slope of the ground, height of the stones, elevation of the horizon, and so on.

Use by Stone Age Man

The most puzzling thing about Callanish is how it was used by Stone Age Britons. I have suggested that Stonehenge was used to mark out the seasons—that the Stonehengers made observations of the moon throughout its 18.61-year cycle in order to establish a lunar-solar calendar and to obtain warning of solar and lunar eclipses. Callanish seems to have been used primarily to establish a calendar, though it may possibly have been used for predicting eclipses as well.

In looking for clues as to how the stones of Callanish were used as a computer to establish a calendar, we find analogies with Stonehenge.

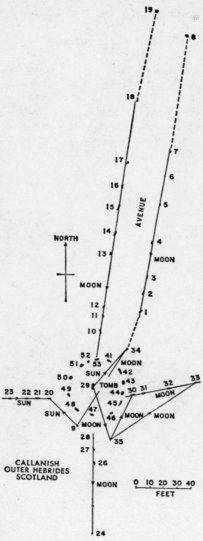

Fig. 2. Plan of Callanish, a group of large standing stones on the island of Lewis in the Outer Hebrides.

Since the circle of stones at Callanish has no solar or lunar alignment I suggest that it is a counting circle similar to the Aubrey holes and Sarsen circle at Stonehenge. The circle at Callanish contains 13 stones, 12 large and 1 small. These numbers are the fundamental basis of a lunar-solar calendar and could have been used for marking off the short years of 12 lunar months and the long years of 13 lunar months. A similar system is still used in the Jewish calendar today. The 19 stones in the avenue, including the "heel" stone (stone 34), provide a basic counting system for this calendar. Such an observational program and calendar formulation in 1500 B.C. would have antedated by more than 1000 years any similar development known to us. The Greek Meton is credited, perhaps apocryphally, with the discovery, in 432 B.C., of the 19-year cycle; this knowledge was not put to use until 312 B.C., during the Seleucid Empire.

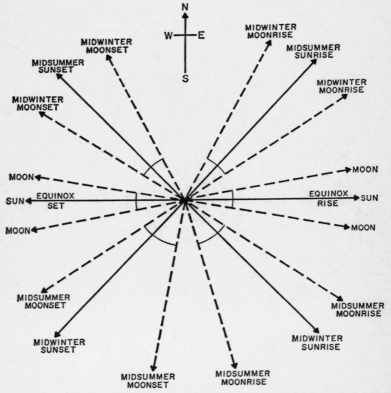

Fig. 3. The azimuthal direction of the rising and setting of the sun and moon at solstice and equinox for the latitude of Callanish.

The Callanish people may have observed and predicted eclipses, though the evidence is less clear than it is at Stonehenge. A midwinter moonrise over stone 34 would certainly have signaled the danger of a winter eclipse. The requirement for winter and summer eclipses is also marked by the lines for moonset and sunrise at the equinox. When the sun rose in line with stones 20 to 23 and the moon set in line with stones 30 to 33, there would have been danger of an eclipse at midsummer or midwinter. Thus the Callanish people did have the means for predicting winter and summer eclipses from observations made at various times throughout the year. However, consistent prediction of the eclipses of a moon of a particular time of year, such as the midwinter moon, would have required a 56-year counting cycle made up of intervals of 19, 19, and 18 years. The Callanish people could readily have made such observations by excluding stone 34 every third count around the avenue. Thus, it is just possible that they did have knowledge of the 56-year cycle, though they did not reveal possession of this knowledge, as the Stonehengers did by setting out a circle with 56 marked points.

Although the astronomical alignments are indisputable, the suggestion of a computer use is, of course, conjectural. At Stonehenge the precise circle of 56 Aubrey holes seems to be connected unambiguously with an accurate eclipse cycle which synchronizes with the year of the seasons. At

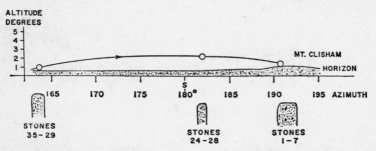

Fig. 4. The apparent path of the full moon at Callanish at midsummer computed for about 1500 B.C.

Callanish, on the other hand, excavations have not been completed. We cannot be sure that only 19 stones were set in the avenue, and that only 13 stones were set in the circle. Also, the circle of standing stones is associated with a tomb and is thought by some archaeologists to be more recent than, and perhaps unconnected with, the rows of stones.

Conclusion

On the basis of the stone record it appears that the Callanish people were as precise as the Stonehengers in setting up their megalithic structure, but not as scientifically advanced. Callanish is, however, a structure that could have been used much as Stonehenge was. It would be interesting to obtain a date, by the radiocarbon method, for the peat in the area of Callanish, to determine how much older, or more recent, than Stonehenge this structure is. Perhaps the knowledge gained at Callanish was later used in the design of Stonehenge.

These structures are both at critical latitudes. Callanish is at the latitude where the moon skims the southern horizon. Stonehenge is at the latitude where at their extreme positions along the horizon the sun and the moon rise at a right angle on the horizon. From the standpoint of astronomical measurement Stonehenge could not have been built further north than Oxford or further south than Bournemouth. Within this narrow belt of latitudes the four station stones make a rectangle. Outside this zone the rectangle would be noticeably distorted. Perhaps these latitudes were deliberately chosen, and perhaps these people were aware that the angles of the quadrangle formed by the station stones would change as one moved north or south. If Stonehenge and Callanish are related, then the builders may have been aware of some of the fundamental facts which served later as the basis of accurate navigation and led to a knowledge of the curvature of the earth. But if they possessed knowledge of such importance it must have been passed along by word of mouth; no record of it is found in the stones.

References

[1] G. S. Hawkins, *Nature* **200**, 306 (1963).
[2] ———, *ibid.* **202**, 1258 (1964).
[3] B. Somerville, *J. Brit. Astron. Assoc.* **23**, 83 (1912).
[4] A. Thom, *Math. Gaz.* **45**, 83 (1961).

BIBLIOGRAPHY

Ancient Irish Tales, edited by Tom Peete Cross and Clark Harris Slover. New York: Henry Holt and Co., 1936.

Antrobus, Lady Florence. *A Sentimental & Practical Guide to Amesbury and Stonehenge*. Amesbury, 1904.

Atkinson, R. J. C. *Stonehenge*. London: Pelican Books, 1960.

Aubrey's Brief Lives, edited by Oliver Lawson Dick. Ann Arbor: University of Michigan Press, 1957.

Barclay, Edgar. *The Ruined Temple Stonehenge*. London: St. Catherine Press, 1911.

———. *Stonehenge and Its Earthworks*. London: D. Nutt, 1926.

Baudoin, Marcel. *La Préhistoire par les Etoiles*. Paris: N. Maloine, 1926.

Browne, Henry. *An Illustration of Stonehenge and Abury*. Salisbury: Printed for J. Browne, 1854.

Butler, Samuel. *Hudibras*. London: Henry Washbourne, 1847.

Caesar, Julius. *Gallic Wars*, translated by H. J. Edwards. The Loeb Classical Library. Cambridge, Mass.: Harvard University Press, 1917.

Camden, William. *Remaines of a Greater Worke Concerning Britaine*. London: Printed by G. E. for Simon Waterson, 1605.

Charleton, Walter, Dr. in Physic. and Physician in Ordinary to His Majesty. *CHOREA GIGANTUM. OR the most famous Antiquity of GREAT-BRITAIN, vulgarly called STONE-HENG, Standing on Salisbury Plain, Restored to the DANES*. London: Henry Herringman, 1663.

Childe, V. Gordon. *Prehistoric Communities of the British Isles*. London and Edinburgh: W. & R. Chambers, Ltd., 1940.

Clapperton, Walter. *Stonehenge Hand-Book*. Salisbury: W. Clapperton, n.d. (about 1850).

Cles-Reden, Sibylle von. *The Realm of the Great Goddess*. Englewood Cliffs, N.J.: Prentice-Hall, Inc., 1962.

Conjectures on that Mysterious Monument of Ancient Art, Stonehenge. Salisbury: J. Easton, 1815.

Cooke, William. *An Enquiry into the Patriarchal and Druidical Religion, Temples &c*. London: Printed for Lockyer Davis at Lord Bacon's Head, 1755.

Creed, Virginia. *All About Ireland.* "The New Europe Guides" series. New York: Duell, Sloan and Pearce, 1951.

Cunnington, M. E. *Woodhenge: A Description.* Devizes: George Simpson & Co., Ltd., 1929.

Cunnington, Robert H. *Stonehenge and Its Date.* London: Methuen & Co., 1935.

The Diary of John Evelyn, edited by E. S. deBeer. Oxford: Clarendon Press, 1955.

The Diary of Samuel Pepys, edited by Henry B. Wheatley. London: G. Bell and Sons Ltd., 1924.

Dillon, Myles. *The Cycles of the Kings.* New York: Oxford University Press, 1946.

Diodorus of Sicily, Book II, translated by C. H. Oldfather. Loeb Library. Cambridge, Mass.: Harvard University Press, 1935.

Edgerton, Harold E. "New light on an Old Riddle," *National Geographic Magazine,* page 846, June 1960.

English Poetry and Prose of the Romantic Movement, selected and edited by George Benjamin Woods. New York: Scott, Foresman and Co., 1929.

Ericson, David B., and Wollin, Goesta. *The Deep and the Past.* New York: Alfred A. Knopf, 1964.

Fergusson, James. *Rude Stone Monuments in All Countries.* London: J. Murray, 1872.

"A FOOL'S Bolt Soon Shott at Stonage," in *Peter Langtoft's Chronicle,* transcribed by Thomas Hearne. Oxford: Printed at the Theatre. 1725.

Geoffrey of Monmouth's Histories of the Kings of Britain, translated by Sebastian Evans. London: Dent. 1904.

Gidley, Lewis. *Stonehenge, Viewed by the Light of Ancient History and Modern Observation.* London: Simkin. Marshall & Co., 1873.

Grover, Henry M. *A Voice from Stonehenge.* London: W. J. Cleaver, 1847.

Hardy, Thomas. *Tess of the D'Urbervilles.* New York: Dodd, Mead, 1960.

Harris, (James) Rendel. *The Builders of Stonehenge.* Cambridge: W. Heffer & Sons, Ltd., 1932.

Herbert, the Hon. Algernon. *Cyclops Christianus; or, an Argument to Disprove the Supposed Antiquity of the Stonehenge and other Megalithic Erections in England and Brittany.* London: Petheram, 1849.

The Historie of the World, Commonly Called, the Naturall Historie of C. Plinius Secundus, translated by Philemon Holland. London: Adam Islip, 1601.

Hoare, Sir Richard Colt. *The Ancient History of . . . Wiltshire.* London: W. Miller, 1812.

Irish Fairy and Folk Tales, edited by W. B. Yeats. Modern Library. New York: Random House, n.d.

James, Henry, Jr. *Transatlantic Sketches*. Boston and New York: Houghton Mifflin and Co., 1903.

James, Sir Henry. *Plans and Photographs of Stonehenge and of Turusachan in the Island of Lewis*. Southampton, 1867.

Jones, Inigo, Esquire, Architect Generall to the late King. *The Most Notable Antiquity of Great Britain vulgarly called Stone-Heng on Salisbury Plain Restored*. London: Printed by James Plesher for Daniel Pakeman at the sign of the Rainbow in Fleetstreet, and Laurence Chapman next door to the Fountain Tavern in the *Strand*, 1661.

Landreth, Helen. *Dear Dark Head*. New York: Whittlesey House, 1936.

Lockyer, J. Norman. *The Dawn of Astronomy*. Cambridge, Mass.: M.I.T. Press, 1964.

——. *Stonehenge and other British Stone Monuments Astronomically Considered*. London: Macmillan, 1906.

Long, William. *Stonehenge and Its Barrows*. Devizes: H. F. and E. Bull, 1876.

Malory, Sir Thomas. *Le Morte D'Arthur*, introduction by Professor Rhys. Everyman's Library series. London: J. M. Dent & Sons Ltd., 1941.

Metrical Chronicle of England, introduction and glossary by Edward Zettl. London: Oxford University Press, 1935.

Modern British Poetry, mid-century edition, edited by Louis Untermeyer. New York: Harcourt, Brace and Co., 1950.

Petrie, W. M. Flinders. *Stonehenge: Plans, Description, and Theories*. London: Edward Stanford, 1880.

Pindar. *Odes*, translated by John Sandys. The Loeb Classical Library. Cambridge, Mass.: Harvard University Press, 1915.

Rolleston, T. W. *Myths and Legends of the Celtic Race*. New York: Thomas Y. Crowell Co., n.d.

Salmon, Thomas Stokes. *Stonehenge, a Prize Poem*. Oxford, 1823 (?).

Selected Letters of Samuel Johnson, introduction by "R.W.C." "The World's Classics" series, Humphrey Milford. London: Oxford University Press, 1925.

Serner, Arvid. *On "Dyss" Burial and Beliefs about the Dead During the Stone Age*. Lund: H. Ohlsson, 1938.

The Works of the Honourable Sir Philip Sidney, Kt. London: Printed for A. Bettesworth, E. Curll, W. Mears and R. Gosling, 1724.

Smith, John, M.D. *Choir Gaur, the Grand Orrery of the Ancient Druids*. Salisbury: Easton, 1771.

Smith, Logan Pearsall. *All Trivia*. London: Constable & Co., Ltd., 1933.

Stevens, Edward T. *Jottings on Some of the Objects of Interest in the Stonehenge Excursions*. London: Simpkin, Marshall and Co., 1882.

Stevens, Frank. *Stonehenge Today and Yesterday*. London: S. Low, Marston & Co., 1916.

Stone, Edward H. *The Stones of Stonehenge; a Full Description.* London: R. Scott, 1924.

Teilhard de Chardin, Pierre. *The Future of Man,* translated by Norman Denny. New York: Harper & Row, 1964.

Thom, Alexander, Professor. "Megalithic Geometry in Standing Stones," *New Scientist,* March 12, 1964.

Van den Bergh, G. *Eclipses,* —1600 *to* —1207. Holland: Tjeenk, Willink and Zoon, 1954.

Webb, John. *A Vindication of Stone-Heng Restored.* London: Printed for G. Conyers, 1725.

White, Gilbert. *The Natural History of Selborne.* London and New York: John Lane, 1902.

Wilde, Lady. *Ancient Legends of Ireland.* London: Ward and Downey, 1887.

Wood, John (Architect). *Choir Gaure, Vulgarly Called Stonehenge, on Salisbury Plain, Described, Restored, and Explained.* Oxford: Printed at the Theatre, 1747.

Young, Ella. *The Tangle-Coated Horse.* New York: Longmans, Green and Co., 1929.

INDEX